BAUDIN, NAPOLEON AND
THE EXPLORATION OF AUSTRALIA

EMPIRES IN PERSPECTIVE

Series Editor: *Durba Ghosh*
Advisory Editor: *Masaie Matsumura*

TITLES IN THIS SERIES

FORTHCOMING TITLES

BAUDIN, NAPOLEON AND
THE EXPLORATION OF AUSTRALIA

BY

Nicole Starbuck

Routledge
Taylor & Francis Group

LONDON AND NEW YORK

First published 2013 by Pickering & Chatto (Publishers) Limited

Published 2016 by Routledge
2 Park Square, Milton Park, Abingdon, Oxfordshire OX14 4RN
711 Third Avenue, New York, NY 10017, USA

First issued in paperback 2015

Routledge is an imprint of the Taylor & Francis Group, an informa business

BRITISH LIBRARY CATALOGUING IN PUBLICATION DATA

Starbuck, Nicole, author.
Baudin, Napoleon and the exploration of Australia. – (Empires in perspective)
1. Baudin, Nicolas, 1754–1803 2. Scientific expeditions – Australia – History –
19th century. 3. French – Australia – Sydney (N.S.W.) – History – 19th century.
4. Australia – Discovery and exploration – French. 5. France – History –
Consulate and First Empire, 1799–1815.
I. Title II. Series
919.4'042-dc23

ISBN-13: 978-1-138-66166-0 (pbk)
ISBN-13: 978-1-8489-3210-4 (hbk)
Typeset by Pickering & Chatto (Publishers) Limited

CONTENTS

ACKNOWLEDGEMENTS

This book has been produced with assistance from a number of individuals and institutions. It began as a doctoral dissertation funded by an Australian Research Council Discovery Project. Supplementary faculty grants and the P. W. Rice Travel Award offered by the University of Adelaide allowed me to conduct vital research and to attend conferences interstate and in France. I thank the librarians at the Bibliothèque Centrale, Muséum Nationale d'Histoire Naturelle, Paris, and in particular, Gabrielle Baglione, curator of the Lesueur Collection at the Muséum d'Histoire Naturelle at Le Havre. I am also indebted to the National Library of Australia, which granted me a Norman McCann Summer Scholarship in 2007 – my gratitude goes particularly to Margy Burn and the librarians of the manuscript and map collections. My thanks must also go to Valerie Sitters at the State Library of South Australia and to Jennifer Genion, research assistant for the Baudin Legacy Project.

For generously sharing their knowledge about science, animals and natural history with me during my doctoral research, I thank Stephanie Pfennigworth and Wolf Mayer, and for their encouragement, support and advice as I turned to 'the book', my deep thanks go to Claire Walker, David Lemmings, Lisa Mansfield, Vesna Drapac, Rachel Ankeny and Shino Konishi. I am profoundly grateful to Jean Fornasicro, John West Sooby and Margaret Sankey for initially inviting me into the world of French exploration and for their dedicated supervision of my PhD project. Jean and John continue to discuss Baudin and his world with me and they have provided invaluable advice and reassurance, however, any faults in this book are entirely my own.

Finally, I owe the most special thanks to my sons James and Hamish and my husband Rob, whose love and patience have enabled me to bring this history to light.

Different versions of chapters 4, 5 and 6 appear elsewhere as 'Nicolas Baudin. La relâche à Sydney et la deuxième campagne du *Géographe*', in M. Jangoux (ed.), *Porté par l'air du temps: les voyages du capitaine Baudin*, special number of *Études sur le 18ème siècle*, 38 (2010), pp. 133–42, 'Neither Civilized nor Savage: The Aborigines of Colonial Port Jackson, through French Eyes, 1802', in A. Cook, N. Curthoys and S. Konishi (eds), *Representing Humanity in the Age of Enlightenment* (London: Pickering & Chatto, forthcoming, 2013) and 'The Colonial Field: Science, Sydney and the Baudin Expedition (1802)', *Explorations*, 52 (June 2012), pp. 3–35.

LIST OF FIGURES

INTRODUCTION: VOYAGING OUT OF THE ENLIGHTENMENT

The ship, for the insider – in all its spaces, in all its relationships, in all its theatre – was always being remade, was always in process. Its story had not ended. The partial history men made of it was always creating something new.

G. Dening

It was early in the spring of 1800 that Nicolas Baudin, accompanied by a commission of the Institut National's most eminent savants, presented the First Consul with his proposal for a voyage to 'interest the whole of Europe'.[1] This was to be his most ambitious project yet: a round-the-world voyage of natural history and geographic research aimed at augmenting 'the discoveries of the great navigators who for over 40 years had radically extended the domains of these two sciences'.[2] In fact, with an itinerary featuring the coastlines of Africa, the Americas, Hawaii, Tonga and Australia, and comprising no less than three ships, it would have been an even greater enterprise than the 1785–8 expedition of François de Galoup de La Pérouse – that expedition designed to rival the Pacific discoveries of James Cook. This was a grand plan, and political instability a year earlier had at the last moment prevented it from being realized, but Baudin had perceived an opportunity in the change of government wrought by the coup of 18 brumaire. He hoped that, if he first guaranteed the support of the Institut, that body 'would engage the government, friend of science' and persuade 'the consuls to accomplish what the others had neglected'.[3] The proposal did immediately capture Bonaparte's interest: 'the reception passed just as I had envisaged', Baudin later noted in his journal, 'the voyage was decided'.[4] Indeed, it was determined that a maritime expedition would take place, but not in the form Baudin had proposed.

Following deliberation by the commission of the Institut, in collaboration with the Minister of Marine and the Colonies and with the formal approval of the First Consul, the voyage that was ultimately decided, the voyage that commenced from the docks at Le Havre, 18 October, little resembled that which had originally been proposed. It involved two ships instead of three but almost three times the number of naturalists and scientists Baudin had requested – an unprecedented twenty-two – and its scope was highly circumscribed: it would be a scientific voyage concentrated on the shores of Australia.

This was a pivotal decision, for it concerned not merely the particular expedition at hand. As they contemplated the merits of Baudin's proposal, the members of the Institut's commission concluded that it was time to institute a change in scientific voyaging. In their report on the inception, preparation and purpose of the Australian voyage, they explained that, after the celebrated expeditions that had criss-crossed the oceans and determined the existence of all the major land-masses of the world, global voyages of discovery were no longer needed. What was required instead were voyages 'restricted to specific, pre-determined, points, directed to the least-known coastlines ... to archipelagos of which the number of islands, their size, their outline and the populations that inhabit them remain to be verified'. These 'direct expeditions', they continued, would also avoid sojourns in well-known regions and instead prolong stays where new geographical observations could be made and new natural history specimens collected.[5]

At a glance, the Institut's resolution may appear to have been merely pragmatic. However, it did take place against a backdrop of Revolutionary transformations in French society, politics and science, and that circumstance begs the question: was this approach about more than matching economic concerns to a desire to fill gaps on the charts of Oceania and in the cabinets of the Muséum d'Histoire Naturelle? In fact, during the course of the Australian voyage, there would be signs of both change and continuity aboard Baudin's ships. However, it is the latter that has been most strongly, and at times nostalgically, emphasized in the historiography of the Baudin expedition. In the late 1960s, John Dunmore declared that the Baudin expedition 'belonged to the tradition of scientific travel that went back to Bougainville; it was the last of the great adventures';[6] and, with recognition of certain developments in natural history,[7] the Baudin expedition has continued to be categorized as an Enlightenment voyage – placed firmly alongside those that had preceded it in the eighteenth century, those that the Institut had decided to leave behind.[8]

Before the historical nature and significance of the Baudin expedition may be reconsidered, then, this category itself needs to be clarified. The Enlightenment was, as John Gascoigne notes, rather an 'attitude of the mind' than a formal doctrine;[9] therefore, it is not easy to define or to ascribe a time-frame. Developing in the seventeenth century and predominating intellectual discourse through the eighteenth century, it is usually considered to have drawn toward a close during the course of the French Revolution and ended, at least for historical purposes, with the Napoleonic era – if not with the establishment of the Consulate then with the creation of the First Empire in 1804.[10] It was an attitude that took inspiration from the ground-breaking advances of the Scientific Revolution and which constituted an embrace of secular philosophy in place of the traditional world view long provided by Christianity. Accordingly, though religion itself was not necessarily rejected, superstition was, and as they pursued an encyclopaedic knowledge and, ultimately, social and political progress, philosophes, men and women in salons and coffee-houses and of course explorers navigating the South Seas put their faith instead in empirical observation.

For the first time in the history of European maritime exploration, the quest for knowledge about the natural world was for navigators an explicit and genuine objective. For the first time too, beginning with Bougainville, they carried naturalists, as well as officers who, like themselves, were trained in natural history and the sciences at naval academies. However, while their explorations across the 'last new world' of the Pacific resulted in the discovery of lands and peoples new to Europeans and fed research in natural history and sciences such as astronomy, there was also a romantic, even utopian, aspect to these voyages. As Dorinda Outram demonstrates, the travel accounts, artefacts and images representing the Pacific world – particularly Tahiti – were taken up by Europeans ultimately less concerned with accuracy than with idealistic projections of their frustrations and aspirations.[11] Furthermore, the culture aboard French discovery ships themselves was naturally characteristic of the *ancien régime* era and carried a sense of grandeur. The commanders and officers, members of the Royal Navy, were all of noble heritage – their authority and rank was justified more by the privilege of birthright than by merit. The expeditions' strategic objectives were vague,[12] though laying claim to new lands was definitely on the agenda. At the same time, the naturalists' research was wide-ranging. Natural history, or natural philosophy as it was most commonly called, was at this stage a sweeping and largely philosophical study of the natural environment: it was implicated in questions about rationality, about the possibility of truly knowing the external world, about how to handle knowledge gathered, and about the 'goodness' of nature versus civilization.[13] Although the study of humanity was not yet a discipline in itself, 'Man' and its relation to the world was central to Enlightenment inquiry. There was confidence that all peoples possessed the capacity for progress and, amongst those such as Jean-Jacques Rousseau and Denis Diderot, the morality of slavery and colonialism was a matter for debate. In all, the Enlightenment introduced a liberating paradigm for intellectual endeavour and opened up the world of Oceania for European exploration. In its application, however, the research carried out by the philosophes and the voyager-naturalists was also marked by uncertainties and contradictions in thought and method.

There would be ambiguities in the practices and theories of Baudin's men too, and philosophical reflections would often frame their observations. Certainly, Enlightenment rhetoric was often applied to the expedition: its purpose as stated by geographer and past Minister of Marine Antoine-Laurent de Fleurieu was 'to increase the mass of human knowledge' while Baudin was even said to be following in the footsteps of 'Bougainville, Cook and La Pérouse'.[14] And there was a curious incongruity in Baudin's itinerary that is indicative of ambiguous strategic interests: a visit to the only colony in the *Terres australes* was not included, still the commander was instructed to report if the British had yet established a settlement on the island of Tasmania. The archives remain silent on whether the exclusion of Port Jackson from the itinerary was due to diplomatic caution, the concern to save time or the expedition's lack of political purpose; no doubt, all three reasons combined played a part.

All the same, the Baudin expedition was clearly distinctive. The commission's plan already indicated a deliberate move toward greater accuracy and detail; its allocation of expert naturalists and artists, a desire for more specialized knowledge and refined skills; its appointment of a merchant-class and self-educated man to the office of command, a commitment to leaving behind the old regime's tradition of grandeur and privileged leadership. Indeed, despite some continuation of the Enlightenment 'attitude of the mind', there were also significant and intended differences that anticipated the more 'scientific' culture of the nineteenth century.

It is largely for this reason that Carol Harrison presents Baudin's voyage, together with that led by d'Entrecasteaux, 1791–4, as a Revolutionary voyage. These were both different from earlier Enlightenment voyages in a number of ways. First, whereas their predecessors had served absolute monarchs, d'Entrecasteaux sailed in the service of the National Assembly and Baudin, as the commemorative medal of his expedition states, served 'Bonaparte, First Consul of the Republic';[15] expeditionary science, Harrison explains, 'had moved from being a project of royal absolutism to a statement of the philosophical ambitions of the revolutionary nation'.[16] Second, both expeditions took place when the prestige of science in the national context reached its height: Frenchmen could make declarations, for example, of 'friendship, science, love and French glory' and there was a strong identification between citizen-scientists, whether in Paris or at sea, and the interests of the nation.[17] Finally, in contrast to the colonizing preoccupation of the British, French national interest in the Pacific was focused on 'scientific investment' secured through French knowledge manifested in maps as well as natural history collections and taxonomies.[18] Neither of these Revolutionary expeditions made any territorial claims; in this period, Harrison points out, France's relationship with the South Seas was crucially different both from what it had been during the Enlightenment and what it would be later in the nineteenth century.[19]

However, while it is entirely appropriate to distinguish the expeditions of d'Entrecasteaux and Baudin, both, from those that came before 1789, it is also worth noting the dissimilarities between them. The d'Entrecasteaux expedition, dispatched in 1791 in search of La Pérouse, carried clear traces of the old order. It was composed of members of the Royal Navy, still in existence, and though many were imbued with democratic principles their loyalties had not yet been truly tested. The d'Entrecasteaux expedition also comprised a scientific staff larger than its predecessors' but much smaller and more traditional than its successor's: three 'naturalists', two astronomers, two hydrographers, two artists and a gardener.[20] Jean-Jacques-Julien de Labillardière's writings about Aboriginal Tasmanians, which of all the outcomes of the expedition carried one of the strongest legacies, conformed closely to the 'noble savage' philosophy popular in the eighteenth century. However, by the time Baudin sailed, the Revolution had reached its height and drawn to a close, the divisive turmoil caused by the fall of the monarchy and the establishment of the Republic – which had torn the d'Entrecasteaux expedition apart in 1793 – had settled down, both naval and scientific institutions had been transformed, and meritocratic principles were being put into practice. The two Revolutionary voyages lay on opposite sides of a moment of rupture: the National Convention's Reign of Terror. As a result, the Baudin expedition was the first to carry to the South Seas a team of naturalists and an officer corps not composed exclusively of aristocratic men, and the first to be led there by a commander not of noble heritage. It was also markedly more scientific in its design than was the d'Entrecasteaux expedition and its voyagers in the field – trained in botany, mineralogy, astronomy, geography, art and medicine – were to demonstrate more scientific methods and attitudes.[21] Indeed, although Baudin's was, at the broadest level, a type of Enlightenment expedition and, more particularly, an outcome of the Revolutionary period, it would seem to have been above all, a Consulate-era voyage – a new voyage for a new nation.

Indeed, the *Géographe* and its consort, the *Naturaliste*, sailed for Australia during a pivotal moment in French history. The nation was moving on from Revolution and toward a relatively egalitarian and democratic yet peaceful and orderly France. Not until the expedition returned would Bonaparte crown himself Emperor; in 1800, he sought to put the philosophies of the Enlightenment and the principles of the Revolution into practice and, by demonstrably doing so, to present himself as an enlightened leader.[22] Baudin had reason to sense that he would embrace his proposal for a scientific venture. Two years earlier he had instigated the celebrated Egyptian expedition and as First Consul he did not hesitate, between implementing administrative reforms in France and leading military battles in Italy, to give his support to an expedition of geographic and natural history research – one not pursuing territorial objectives but certainly promising knowledge of national and political value.[23]

This is not to suggest that the Baudin expedition was a personal project of Bonaparte's. As demonstrated, it was in fact instigated by Baudin himself and organized by the Institut; moreover, it is true that, while it was approved and given funding by the First Consul, the archives give no indication that he traced its progress on charts at his desk or inquired regularly about its accomplishments.

It is no doubt largely for this reason that the expedition is seldom mentioned in histories of the Napoleonic era and, conversely, that on the rare occasions it is mentioned, as if it makes little sense otherwise, it is put forward as an example of the expansiveness of Bonaparte's colonial ambitions.[24] Stuart Wolf, in passing, takes the most nuanced approach when he notes that the Baudin expedition followed the example set by La Pérouse but was also 'the ideal continuation' of the Egyptian expedition.[25] In fact, as a mission supported by the First Consul but not intended for territorial gain or military conquest, the Baudin expedition demands broader recognition: it was the only scientific voyage of the Napoleonic era – from the beginning of the Consulate in 1799 to the restoration of the Bourbon monarchy in 1815 – the period that marked the closing of the Revolution and the consolidation of a new nation, and the fading of the Enlightenment into the nineteenth century.

The most critical feature of this period in relation to the Baudin expedition was the transformation of natural history. This was a process that had of course been occurring gradually but it was markedly accelerated by Revolutionary changes and democratic concerns and it reached a noticeably transitional phase at the turn of the nineteenth century.[26] Indeed, if, as Harrison observes, naturalists had gained status during the 1790s by staking the significance of their work to the interests of the nation, it was from the period of the Directory through to the beginning of the First Empire that science actually became integral to the consolidation of Revolutionary change in France. To apply democratic principles, to establish order and strengthen their authority, to make 'La République, une et indivisible' into an effective and ongoing reality, government administrators pursued with urgency a thorough and precise knowledge of the nation's people and resources, and of the nature of humanity and the natural world in general. As Claude Blanckaert notes, the voice of reason no longer sufficed,[27] the methods needed to be more systematic and precise, and the knowledge was not to be merely 'encyclopaedic', but concerned particularly with social and political progress. This was what Blanckaert calls, writing particularly in relation to the 'science of man', 'le moment naturaliste', and it was generated not only by the demand for knowledge but also by the reorganization of scientific societies and institutions that had begun in the mid-1790s.[28] Previously accessible mainly to the privileged, major institutions such as the Muséum d'Histoire Naturelle in Paris became state-funded and staffed by professional naturalists. The broad domain of 'natural history' steadily separated into a variety of sub-disciplines and an increasing distinction was drawn between research carried out in the field and analysis undertaken in the *cabinets* of the Museum. Of course, not all naturalists moved forward at the same pace – naturalist-explorer Alexander von Humboldt, for instance, revelled in forays into the Americas while continuing to advance knowledge across disciplines – and naturalists in the field did not always carry out their research quite as the innovators in the metropolitan centre had anticipated; nonetheless, this was unquestionably a distinct and crucial turning point in the history of science and, with all its ambiguities, that evolution was bound to affect the nature of scientific voyaging.

Accordingly, from its initial conception in Paris down to its performance in the *Terre australes* – particularly its performance – this expedition demands a fresh investigation. An exceptional opportunity to this end is provided by a particular stage of Baudin's voyage: the unscheduled and prolonged sojourn in the nascent British penal colony at Port Jackson (known today as Sydney Harbour), in 1802. The colony at this stage was in only its fifteenth year. La Pérouse had visited nearby Botany Bay in 1788, just days after the arrival of the First Fleet, but Baudin's was the first French expedition to visit the established colony at Port Jackson. During these five months at anchor, at the furthest point of its journey, the expedition's culture and mechanics were laid bare. The voyagers socialized and exchanged knowledge with English settlers, merchants, naturalists and explorers and day-to-day they observed the local Aboriginal people dealing with colonization. The naturalists journeyed into the interior to observe the natural environment and to collect specimens and, at other times, indulged in the opportunity to barter for objects and to study their collections in Sydney. The naval staff, for their part, maintained the ships while they lay at anchor and prepared them for their return to sea. And, notably, the commander, while supervising these activities and managing diplomatic relations with the Governor of New South Wales, reconstructed the expedition. Having already surveyed the western, Tasmanian and southern coasts of Australia as instructed and with only the north-west coast yet unseen, he sought to recommence his exploration of Australia and satisfy more fully the aspirations of the Institut and the Consulate. He returned the *Naturaliste* directly to France, carrying the immense natural history collection that had been gathered by that point and the men whose health or conduct he deemed disadvantageous to his new expedition, and he replaced it with a new, smaller vessel: the *Casuarina*, which would be used to achieve more accurate coastal examinations. Finally, based on the knowledge he understood was most vital to the Institut and the Consulate at this time, he carefully revised his itinerary and his approach.

In effect, while the nation of France was undergoing a transitional period under the leadership of Bonaparte, so, on the other side of the world, this scientific expedition was undergoing a transition of its own. The conclusion of one voyage at Port Jackson and the commencement of a new one affected a liminal space between voyages: the possibility of ending one's mission and returning home or of embarking on a second campaign in order to achieve greater results proved both confronting and liberating for Baudin and his men. They were led to reassess, reengage with and reveal their most fundamental values and aspirations, to decide where they stood concerning their presence as French voyagers in the British colony. How did the Frenchmen identify themselves in their interactions with each other, with the English and the Aboriginal people? What views about human nature and human rights did they express, perhaps unwittingly? How did they approach and give meaning to their natural history research and to the objectives of the expedition overall? The commission of the Institut had designed this venture explicitly as a new style of scientific voyage – how new was it in the end? And correspondingly, from below to quarter deck, from the portside in Sydney Town to the foot of the Blue Mountains, what precisely did it reveal about the world from which it was constructed?

It is by connecting the over-arching interests of the French nation, government and science with the everyday gestures, declarations and decisions of the commander and his men at Port Jackson that such questions may most effectively be elucidated. Accordingly, while the dual transitional nature of the sojourn overarches this history, there are three central themes underlying it. The first concerns the spaces in which Baudin and his men collected and observed, socialized and negotiated, and re-evaluated the purpose of the expedition. With its townships and its farmland, its harbour and 'bush' – particularly with its history of attempts to rehabilitate convicts, 'civilize' Aboriginal people and bring 'order' to the Indigenous landscape – the colonial space functioned both as a 'Pacific theatre', where the voyagers acted out their imperial, scientific and masculine selves,[29] and as a type of 'field' in which the naturalists' research was shaped by access to transport and accommodation and by the sharing and politics of knowledge.[30] Of course, the world of the ship was here too, a little less confined and differently ordered than when at sea. Altogether, it was a rich and complex terrain that evoked and often challenged French ideas on a range of contemporary issues, particularly colonialism, humanity, class and natural history, as well as the nature of scientific voyaging. The most crucial part of this space was its inhabitants; and it is the voyagers' contact with the British that constitutes the second central theme. The occurrence of a French sojourn on His Majesty's territory did not come as easily – and was certainly not so simply attributable to British generosity – as previous histories have suggested.[31] The accommodation, the opportunities for research, the resolution of conflict and the means to prepare a new expedition were accomplished largely through negotiations and diplomatic exchanges. It is in these interactions that one may glimpse the Frenchmen's sense of manly honour and patriotism as well as of pride in the purposes of their expedition. The influence of contact with Port Jackson's Aboriginal people, by comparison, left only a shadowy trace in the expedition's records, which in itself reveals something deeper about the voyagers' feelings on human nature and perhaps the project of 'civilization'. Naturally, such attitudes, like the newly formed relationships and the responses to the colonial environment, did vary from one voyager to the next. Accordingly, the third thread running through this history is the significance of individual agency; that is, the differences in attitudes, behaviours and decisions related largely to social background, rank and experience. As Nicholas Thomas points out, just as we try not to stereotype the Indigenous people they encountered on their travels we must also avoid stereotyping the European voyagers themselves.[32] The scientific and imperial interests of the French nation did not arrive unadulterated on Port Jackson's shore. They were performed by sailors, officers, naturalists and a commander, who each responded in their own way to the physical and implicit features of the colonial environment.

While the main focus here is the everyday detail of individuals, contact and spaces at Port Jackson, 1802, this volume necessarily begins and ends in Paris. From Bonaparte's leadership as First Consul, chapter 1 establishes the background of the men and the decisions that initially shaped the Baudin expedition and would later influence the voyagers' conduct in port. It then traces the conception of the expedition and Baudin's decision to diverge from his itinerary and spend the winter at Port Jackson. Thereafter, beginning with a detailed setting of the scene in chapter 2, it is the men's activities in the colony that take centre stage; however, the recent history of the French navy, Anglo-French relations and ideas about and approaches to studying humanity and natural history remain in the background.

Chapters 2 to 5 concern the Frenchmen's relationships – not only on shore with the people of Port Jackson, but also aboard the *Géographe* with each other. In fact, a particularly interesting, though little-studied, cultural history attaches to the expedition's naval staff. As noted, Baudin was the first *officier bleu* to command an expedition to the Pacific and the first to lead sailors and officers of the Republic in these waters. Chapter 3 considers the conduct of these men as they remained at anchor for five months in the British port: how they handled issues of class, politics and rank, how they related to the British officers on shore and how they felt about their role in the mission at hand. Furthermore, it is not forgotten that Baudin and his men had spent much of their careers in battles against the British, that France and Britain had in fact still been at war when the expedition left Le Havre and that, to make matters worse, there was a history of Anglo-French rivalry in Oceania that was far from ending. While the 'sciences' may never have actually 'been at war',[33] the expedition's relationship with the British colonists – their hosts, rivals and intermediaries – is clearly deserving of attention. Beginning with the involvement of Sir Joseph Banks – unofficial head of all business relating to New South Wales – chapter 4 thus looks at the Frenchmen's relations with Governor King above all, but also with the officers of the New South Wales Corps and with the Flinders expedition, which was also in Australian waters at this time. By comparison, French contact with the Aboriginal people of Port Jackson had less impact on the expedition but, in terms of what it reveals about the voyagers, it was nonetheless significant. This sojourn took place at the end of a revolution concerned with *liberté, egalité et fraternité* and during what has been identified as a period of rapid development in the natural history of man. Giving priority to these circumstances, chapter 5 analyses the French accounts of Aborigines in a colonial situation and considers what these reveal about the authors' ideas of humanity and 'civilization'.

Chapters 6 and 7 shift the focus to the practices of voyager-naturalists in the field and the practice of scientific voyaging itself. Linking to the previous chapter, the collection of ethnographic objects is discussed in chapter 6, in the context of the diverse studies in natural history and sciences undertaken during these months. Ten young naturalists remained with Baudin at Port Jackson, all recently trained in their respective disciplines. Their work, as well as Baudin's, is examined in this chapter with a view to how, in this colonial field, they each gathered, organized and gave meaning to their findings. Of course, Baudin played a central role in each element of the sojourn and his reconstruction of the expedition affected his men in the field, aboard the ships and in their relations with their British hosts. However, exactly how he went about preparing a new voyage, what he was trying to achieve and what that voyage was like is explored in-depth in chapter 7. It is this new voyage that completes Baudin's exploration of Australia and returns the *Géographe* to the eve-of-empire nation of 1804.

France had altered significantly in the four years that the expedition had been at sea. That moment in Bonaparte's rule when such a voyage as Baudin's had been possible, even desirable, had passed. Yet, for its brevity and its pivotal nature, that moment is all the more intriguing and the slice of it that is encapsulated in the Australian expedition, all the more valuable. The five months that this expedition spent between voyages, its purpose and design being re-evaluated, has much to reveal: how was post-Revolutionary, end-of-Enlightenment France played out on the shores of Sydney Town? And in line with the contemporaneous transformation of French society and government, did this expedition, for Baudin as well as for the Institut National, truly represent a transitional point in the history of French scientific voyaging?

1 BETWEEN REVOLUTION AND EMPIRE: FRANCE AND ITS AUSTRALIAN VOYAGE IN 1800

The *Géographe* and the *Naturaliste* embarked men from each rank of French society, from the navy, the army and the sciences, from across the Republic. During their four years at sea their decks hummed with talk – whispered wistfully between hammocks at night or shouted in anger at table in the great cabin – coloured with memories of revolution and war, with dreams of new lands and new peoples. Indeed, while France was advancing toward the era of the Grand Empire, Baudin's ships continued to carry the varied yet characteristic concerns, aspirations and ideals of 1800; and, eventually in Sydney, the voyagers would, each in their own way, act out their post-Revolutionary status. Baudin himself would try to evoke the ambitions of his nation as he endeavoured to prepare the perfect voyage for his superiors. However, performing and fulfilling the desires of turn-of-the-century France was not a straightforward matter. This was a period of transformation in French society and government: the end of the Revolution and the beginning of the 'Napoleonic adventure',[1] the promise of stability following upheaval, the waning of Enlightenment philosophy and the rise of professionalized science. Therefore, in order to understand the expedition's sojourn at Port Jackson, consideration must first be given to the key features of its background: the complex and changing nature of early Consulate-era France, the emergence from this world of the Baudin expedition, and how – in setting sail for Port Jackson – that expedition eventually diverged from its official instructions of 1800 in order to more satisfactorily complete them.

Early Consulate-Era France

Bonaparte had assumed the role of First Consul only eleven months before the Baudin expedition set out for Australia; even so, his influence in political circles, on the position of France in Europe and on the *mentalité* of the French people had already proven potent.[2] During the mid-1790s, Bonaparte's triumphs in the field had corresponded fortuitously with the growing militarization of France and a growing need among the people, including the revolutionaries, for a more robust and centralized government. With each victory, Bonaparte's authority and ambition had increased. Upon his return from Italy in 1797, Parisian crowds had greeted the *petit caporal* as a hero: 'the victor of Italy, the peacemaker of Campo Formio';[3] then, in 1798, he launched the Egyptian campaign. The thought of striking at the heart of British imperial power while extending French civilization to the Orient appealed to the Directory and delighted the people. Although he returned to France in 1799 from a disastrous battle, both the Corsican general and his campaign continued to symbolize French glory and power and Bonaparte was once more embraced as a hero and a saviour. The political coup of 18 November 1799, which, with the support of the *ideologues*,[4] finally replaced the Directory with the Consulate and established Bonaparte as the First Consul, was observed with satisfaction by the vast majority of French people. The people's optimistic devotion to the new First Consul had merged with an increasingly optimistic view of the nation and, ultimately, raised French patriotism to new heights.

Bonaparte immediately and firmly took the nation into hand. He created a new political system which would eventually end parliamentary rule and put himself above his colleagues in terms of accountability and power; still, he was seen to be fulfilling his promises to the people of France.[5] He established a new, more centralized, bureaucratic system which considerably strengthened executive power. It gave the First Consul direct responsibility for appointing ministers and officials,[6] and enabled him to initiate legislation after consultation only with a Council of State appointed by himself.[7] Departmental prefects and propertied local councils were given greater authority over small provincial communities, but Bonaparte did not intend to allow any part of France to escape the attention of the broader bureaucratic body or of himself. Following the example set by the Directory, he and his officials sought to facilitate their tight administration of France via a comprehensive statistical knowledge of its resources. Early in the era of the Consulate statisticians threw themselves, 'in a burst of anthropological curiosity', into the gathering of existing and new data to introduce national registers of property and persons.[8] Overall, this scientific administration emphasized the importance of order.

The Revolution had thoroughly disrupted France's economy as well as the nature of its social hierarchy. Family fortunes and positions were lost through the abolition of venal offices and the devaluation of government bonds, merchants were hit by the collapse of the luxury trades, while employees of government agencies lost their positions. The Revolutionary Wars had affected the economy broadly, as indeed, of course, had the emigration of hundreds of the nation's wealthiest consumers.[9] However, some people actually benefitted from the changes wrought by the Revolution: for example, Revolutionary institutions and the armaments industry provided new employment opportunities in Paris, the abolition of the trades corporations allowed artisans to cross trade boundaries, the property market picked up and sectors such as higher education and the publishing industry even prospered.[10] In 1800, the French were poorer than they had been prior to the Revolution, but the economy was clearly recovering and in Paris, as no doubt elsewhere in France, the social elite was gradually being re-established according to wealth rather than birth.[11]

Also underway by the turn of the nineteenth century was the recovery of the Catholic Church. In 1792, Revolutionaries had closed down convents and monasteries across France, required priests to join the civil service by swearing an oath to liberty and equality, and either executed or forced into exile those who resisted. The closure of churches had soon followed and, in some regions, Catholic practices had been prohibited. On the surface at least, French society was quite thoroughly secularized. In 1795 the Directory reopened churches and, in 1796, allowed priests to return if they agreed to take the civic oath, however, there were still to be no public displays of religion and the separation of Church and State was maintained. Bonaparte was eager to continue this arrangement but also to officially re-establish the Church: 'the people need a religion', he pronounced, and 'this religion must be in the hands of the state'.[12] Even before the Baudin expedition had set sail, the First Consul had already made advances to the Pope.

The Church's influence on society, however, was not to be as pervasive as it had been during the *ancien régime* as it was no longer responsible for the provision of major services and institutions – poor relief, hospitals and, in part, education had all been taken over by the state. The young adults of 1800 may have been schooled in their early days by clerics – though many nobles were educated privately – but those who pursued higher education attended secular institutions such as the École polytechnique, the École des mines and the Muséum National d'Histoire Naturelle, all established during the 1790s. As Martyn Lyons points out, these institutions reflected the importance that the Napoleonic state placed on scientific knowledge and, too, the gradual professionalization of the sciences that occurred over these years.[13] It was in fact establishments such as these that produced the French doctors, engineers and naturalists of various disciplines – including voyager-naturalists – of the nineteenth century.

Of course, these were opportunities available exclusively to male students. Bonaparte declared that girls 'cannot be brought up better than by their mothers. Public education does not suit them, as they are not called into public life; manners are everything for them; marriage is their whole destination'.[14] This was not a new idea, during the Enlightenment the notion that women were naturally suited to domestic and not public life had been powerfully promoted by popular works such as Jean-Jacques Rousseau's *Emile*. However, during the Revolution, secularization as well as the ideals of liberty and equality afforded women increased rights, most notably around inheritance, divorce and birth control. As Olwen Hufton remarks, *Emile*'s 'Sophie' was by and large a peacetime ideal.[15] In 1800, the rights granted to women during the Revolution were still in place but the idealization of the patriarchal family unit was being explicitly reinforced.

Moreover, at the same time that women were romantically portrayed as, and in fact by law forced to be, reliant on their husbands, fathers or brothers, Revolutionary and Napoleonic developments frequently and for long periods took their men away from them. For example, by the time that Bonaparte had seized power, over one million Frenchmen between the ages of twenty and twenty-five had been conscripted to serve in the army.[16] A profound concern for many of Baudin's men, as they prepared to set sail for the South Seas, was how their families would survive in their absence. The astronomer Frédéric de Bissy was only one of the voyagers to appeal to the government for assistance: 'I could not leave my wife and daughter without first ensuring their survival', he wrote to the Minister of Marine and the Colonies. He requested that the government provide his family with a regular income during the course of his absence, remarking: 'you understand too well that it is the responsibility of a Father and a Husband, to refuse me this justice'.[17] Such requests were evidently standard practice in the navy, for the letter written by Bissy's colleague Bory de Saint Vincent is composed in the same terms; still, that is not necessarily to suggest that the emotion and sense of responsibility inherent in these appeals were not genuine. Indeed, it was so that he could care for his elderly mother that, just as the *Géographe* and the *Naturaliste* were about to depart from Le Havre, Baudin's senior botanist, André-Pierre Ledru, withdrew from the expedition. A scattering of correspondence clearly shows that several wives, mothers and entire families were left in a precarious position by the departure of Baudin's expedition. In the voyagers' absence, the gendered restrictions of the Napoleonic Code were to limit women's self-sufficiency even further.

In fact, perhaps more than ever, women were playing a valued but secondary role not only to men but also to France – Napoleonic culture was unambiguously masculine. It encouraged Frenchmen to be authoritative, responsible and virile husbands, certainly, but above all to stride out into the world as representatives and agents of a great nation. The epitome of this type of masculinity was the soldier. Bonaparte had begun actively to propagate martial virtues in France almost immediately upon becoming First Consul – his establishment of the *lycées* was one line of attack to this end, the militarization of masculinity, another, more pervasive, method. As Michael Hughes explains, at the core of this Napoleonic manliness laid two fundamental attributes: honour, the drive to earn respect as well as personal distinctions and privileges, and the pursuit of glory, gained through exceptional achievements. Thus it could be fulfilled not only on the battlefield but also through artistic and intellectual endeavour.[18] It was most certainly seen to be demonstrated through such endeavours as those to be undertaken by Baudin and his men: Louis-François Jauffret, for example, referred to the voyagers enthusiastically as 'courageous men who go to face such dangers in order to increase our knowledge!'[19]

At the turn of the century there was a feeling in France, particularly among its more comfortable and prosperous citizens, of not only stability and peace but also triumph. It was a sense that the French nation had been revolutionized, rejuvenated and had achieved a heightened pre-eminence. The regime, with an orderly social base and military and intellectual support, was poised to renew its position in the world by both enriching and expanding French civilization.

France, Empire and the World

Indeed, by the hearth, in the office, on the battlefield, in the colonies and at sea, Frenchmen were expected by the Consulate to play a leading role on the global stage. By no means had the turmoil and transformations of the 1790s or Bonaparte's efforts to bring order to the Republic led French society to turn inward. France at this time was composed not just of those regions and peoples within its hexagonal borders, but also of French possessions in the Caribbean, South America, the Indian Ocean and Africa. This diversity and global spread was reflected in the composition of the Baudin expedition. Baudin's men came mainly from Normandy, Brittany, northern France and the Atlantic coast; however, especially after others embarked during ports of call, the expedition also included men from Saint-Domingue, Guadeloupe, Spain, Africa, India and the Isle de France.[20] They represented a geographical mobility that facilitated French prominence in the Republic of Letters and an empire that was vital to the geo-political status and economic prosperity of France. Colonial ventures, for instance, were the source of one-third of French trade,[21] and the colony of Saint-Domingue – which in 1800 was being wrenched from France by rebellious slaves under Toussaint L'Ouverture – alone produced almost twice as much revenue in exports as its neighbouring British colonies.

In varied measures, success in these imperial ventures typically involved the combined interests and efforts of the government, the scientific establishment and the navy, and this interdependence was distinctly reflected in the nature of the French navy itself and particularly in its transformation during the second half of the eighteenth century. Naval men had always possessed a degree of worldliness and certainly considerable skills in areas such as hydrography and astronomy, simply by virtue of the extended periods they spent at sea; however, in line with the Age of Enlightenment, this period saw a more serious spirit of inquiry develop among the officers of the French navy than ever before. Naval authorities, for example via the establishment of the Academie Royale de la Marine at Brest in 1752, actively sought to develop scientific expertise in officers that could be applied to exploration. As this approach continued into the 1790s, however, the entire fleet was otherwise transformed as the ideology of the new Republic was imposed upon every rank. Below deck, chaplains were replaced by instructors, charged with providing a basic education and teaching revolutionary principles to naval crews, while in the great cabins or on the docks, officers were screened for disloyalty to the new regime.[22] Traditional practices may not have ceased altogether – in 1800, certain of Baudin's junior officers were recruited at the request of noble families – yet meritocratic principles certainly took effect. Indeed, it is almost certain that had it not been for the revolutionization of the navy, Baudin himself would not have stepped ashore in Australia as commander of the expedition.

The reforms, however, had also caused considerable disruption. During the mid-1790s, at the height of the Revolution, naval crews and officers went to sea highly politicized and naval operations were impeded by a series of mutinies. It was not until the overlapping eras of the Directory and of Bonaparte's rise to power that a degree of order and sense of purpose gradually began to develop in the navy. During the Irish expedition of 1796 and the Egyptian campaign of 1798, the sailors' republican fervour was brought to the traditional preoccupation of the French military: battling the British. Still, success in battle proved elusive. Hopes of conquering the Orient and the Indies, for instance, were swiftly crushed by what would become one of the French navy's most infamous defeats at the Battle of the Nile. All in all, by 1800, the new Republican navy was finding its feet, but morale remained low.

At the same time, war and revolution had had a momentous impact on the scientific establishment of France. During the Revolutionary and Napoleonic wars, armies and, in the case of the Egyptian campaign, accompanying savants had gathered curiosities and natural history specimens from subjugated nations, considerably augmenting the collections held at the Muséum d'Histoire Naturelle in Paris. The Egyptian expedition had been spectacularly successful; indeed, it was that, rather than the military defeat, that captured the interest of the French public. Almost 150 naturalists had accompanied Bonaparte on his campaign, entrusted with the comprehensive study of Egypt's natural, exotic and cultural features. They established the *Institut d'Egypte* in 1798 and continued their research there until the end of the campaign in 1801. At the same time, the Revolution had given rise to the transformation of the Paris Muséum itself. Originally the Jardin du Roi, a medicinal garden maintained for the purposes of the king, it became a national establishment belonging to the public. Here was the beginning of a reconceptualization and reorganization of the institution: in 1793, the National Convention officially founded the Muséum National d'Histoire Naturelle with a new democratic constitution, a new egalitarian administration and the establishment of twelve disciplinary chairs. This was a defining moment in the transformation of French natural history – in Claude Blanckaert's words: 'the transition between the *Ancien Régime* of the sciences and the intellectual revolution triggered by the advent of natural history disciplines'.[23]

With this development, naturalists were ever more explicitly making the globe their laboratory and, in order to advance their production of global knowledge, distinguishing ever more distinctly between sedentary and field research. Certain naturalists, such as Bernard Germain de Lacepède, continued well into the nineteenth century to take a broad-spectrum and philosophical approach to natural history.[24] However, a new generation of French naturalists was progressively narrowing and deepening its research, moving away from the Enlightenment pursuit of encyclopaedic knowledge and absorption in natural philosophy to specialized analyses – particularly in the biological disciplines – that were intended to found new scientific theories. These sedentary naturalists inhabited the physical world of the museum but, as comparative anatomist George Cuvier argued in 1807, intellectually they 'roam[ed] freely throughout the universe' via leisurely study of the products field naturalists had laid before them.[25] It was thus the field naturalist's duty to focus more on collecting specimens and information, according to given instructions, than on analysis. While the sedentary naturalist therefore carried more scientific authority, it was the field naturalist – popularly imagined as 'struggling over remote and dangerous terrain in dedicated pursuit of new and strange plants and animals' – who was held up and romanticized as heroic and manly.[26] The division of course was not absolute, eminent members of the Institut National accompanied Bonaparte on the Egyptian Campaign and several of Baudin's naturalists were highly educated and aspiring to excel in specific disciplines. Nonetheless, the 'sedentarization of knowledge'[27] was developing rapidly and expeditions were seen increasingly as instruments for natural history collecting, designed to feed metropolitan science.

The modernization of French natural history is exemplified particularly well by the natural history of man. As Claude Blanckaert explains, the study of human nature, of human progress and natural rights, had taken on a heightened urgency since the height of the Revolution.[28] Under the Directory and then under Bonaparte, administrators and naturalists sought a more detailed and profound knowledge of humanity in order to understand how far democratic principles could actually be applied and to facilitate the government's efforts to unite the French as 'une et indivisible'. They were therefore interested not only in the distant Other, the Indigenous peoples of Oceania, Africa and the Americas for instance, but also the African slaves in their colonies – freed in 1793 – as well as the 'savages' of rural France itself. The concept of 'race' was not yet defined precisely in its modern biological sense. Republican ideals encouraged naturalists to look more for similarity than for difference.[29] However, as research grew increasingly specialized, as the natural history of man drew increasingly on the biological sciences, and as interest in the nature of 'man' heightened, the questions that many naturalists posed and the methods they developed became respectively more penetrating and more systematic than they had ever been before. Philosopher Joseph-Marie Degérando continued to concentrate on cultural rather than physical difference and to explain such difference in terms of degrees of civilization; however, the method of field observation he advocated was rigorous and, above all, objective. At the same time, comparative anatomist Georges Cuvier sought answers by connecting physiological features – particularly cranial dimensions and facial angles – to moral and intellectual characteristics.[30] For the time being, it was determining how to 'stablize society' and to guarantee individual rights that, as Martin Staum states, was of fundamental importance,[31] but the stage was being set for the development of physical anthropology in the nineteenth century and the nature of 'man' and 'civilization' was a matter for fierce debate.

Indeed, this and many other problems of natural history were discussed widely not only within the grounds of the Muséum but also between correspondents of the commonwealth of learning and among the members of organized intellectual societies. One of the most interesting of these societies was the Société des Observateurs de l'Homme. Taking advantage of an exceptional event, the discovery of the 'savage child of Aveyron', it was established in 1799 by Louis-François Jauffret. It aimed to 'perfect the science of man' through research in the physical, intellectual and moral natures of the human – thus ambitiously initiating a science defined by Jauffret as 'anthropology'.[32] The Society attracted the membership of Degérando as well as Nicolas Baudin and a number of his men but, dissolving in 1804 – the same year that the Australian voyage was concluded, it was short-lived. Supposedly, certain members of the expedition also belonged to a far more long-standing association: the Freemasons. Freemasonry had begun in Britain and, from there, cultural values such as religious toleration, democratic government, egalitarianism and progress based on merit had spread to continental lodges during the eighteenth century. Not surprisingly, the fraternity was blamed by some for the instigation of the French Revolution, but it survived nonetheless and remained popular with Frenchmen of the middle and aristocratic classes.[33] However, the largest and most influential association of this era was the Institut National de France, established in 1795 to replace the Royal Academies abolished earlier by the National Convention. The Institute was intended to be 'the epitome of the learned world, the representative body of the Republic of Letters'.[34] In fact, it brought together Europe's intellectual elite, more than half of whom represented the sciences and several of whom also held positions in government, and thus wielded considerable power in France. Bonaparte himself was elected to the Class of Physical and Mathematical Sciences in 1797.

Intellectual accomplishments during this era were deemed to involve not merely the advancement of human knowledge itself but also the morale and repute of the French people. Accordingly, the steady stream of expeditions sent from France to the Pacific since the mid-eighteenth century continued even during the Revolution. In 1791, during the era of the constitutional monarchy, the Société d'Histoire Naturelle successfully petitioned the National Assembly to send a discovery voyage to the South Seas. Bruni d'Entrecasteaux was to carry out scientific and commercial investigations as well as to search for the missing expedition captained by La Pérouse, last seen at Botany Bay in 1788. Such was the French peoples' concern for the lost voyagers, so the story goes, that King Louis XVI inquired at his execution: 'is there any news of La Pérouse?' Whether these were his last words or it is simply a myth, the story nonetheless illustrates the fact that the pursuit and assertion of human knowledge was tied just as tightly to national pride during the Revolution as at any other time in French history, if not more. In fact, after the disintegration of the d'Entrecasteaux expedition in 1793 and the subsequent appropriation of its natural history collection and papers by the English, not only the botanist Labillardière but also the 'legitimate' king of France, Louis XVIII, and the Directory all mounted a vigorous campaign to recover the results of the voyage. Their efforts continued for some years but, in the meantime, the Republic dispatched another, more modest, scientific mission: Nicolas Baudin's botanical voyage to the Caribbean, aboard the *Belle Angelique*. Baudin had spent the late 1780s and the initial years of the Revolution transporting botanical collections for Austria before returning to the service of his own nation. In 1796, the Institut and the government assented to his proposal to retrieve a botanical collection he had left, following a shipwreck, in Trinidad. Upon the expedition's return in 1798, exotic specimens that Baudin and his naturalists had collected – banana and coconut palms, pawpaw trees – were included in the grand parade of the fête de la Liberté, 1798, alongside the spoils from Bonaparte's Italian campaign. Thus, on the Champ de Mars that year Parisians gathered in jubilant celebration not only of 'liberty' but also of empire and discovery.

At the turn of the century France's relationship with the wider world was, like its internal state, multifaceted and evolving rapidly. Many and various minds – analytical, philosophical, political – were put to the task of planning Baudin's expedition; eventually, diverse interests would guide it along the coasts of Australia and shape its representation in Sydney Town. Indeed, a range of experiences, of the Republic at home and abroad, were brought to this voyage. Some of the expeditioners had taken part in earlier Revolutionary voyages – Baudin had captained two; many of the men had spent years fighting the British at sea – François Péron and midshipman Joseph Ransonnet had fought on the ground in the Revolutionary army; and, a number had passed the mid-1790s studying in Paris. Altogether, the preparation and composition of this expedition would inevitably manifest the competing continuities and changes of the post-Revolutionary nation.

A Voyage in the Time of Bonaparte

Baudin's proposal for a global voyage of discovery had been conceived almost as soon as the *Belle Angelique* was anchored in 1798, just shortly before Bonaparte embarked on his Egyptian campaign. While the General had been occupied in the Orient, Baudin had been developing his plan, refining his objectives and securing the support of the Institut National and the French government. It will be recalled that this grand expedition was approved by the Directory and that, in fact, it had almost gone ahead: as Baudin had been about to set sail in the spring of 1799, the government postponed 'this beautiful expedition for an indeterminate period'.[35] Remarkably, it was only one year later that the new government, headed by Bonaparte as First Consul, swiftly agreed to support the same proposal; though, as highlighted earlier, that plan was significantly altered – Baudin's expedition was made at once more geographically circumscribed and more scientifically ambitious.

It is important to note that this new plan for a *voyage aux Terres australes* was an extension of what Baudin himself had highlighted as 'one of the expedition's most important objectives: a detailed examination of the south-west part of this large island which might perhaps form a continent'.[36] There were new discoveries to be made on 'New Holland's' south coast, he declared, that could advance human knowledge across the fields of natural science and benefit the nation's political economy. In fact, the commission's proposal and the official itinerary laid out by the Comte de Fleurieu both distinctly echoed these assertions and expanded upon them. They directed Baudin to take particular care in charting the south-east coast but also to extend his exploration to the western, northern and Tasmanian coastlines. Moreover, they passed over the commercial objectives emphasized in the captain's own proposal and stressed instead the benefit for the progress of science as well as, via knowledge of the Bass and Torres Straits, for strategic knowledge.[37] For its promises to French science and power, *la terre Australe* was thus singled out as the most important region for Consulate-era France to reconnoitre.

In the hope of realizing these promises, the Institut National worked to assemble a unique balance of naval and scientific staff. Led by the largest team of naturalists ever to take to the seas, in collaboration with and steered by a naval crew, the Baudin expedition was to be truly a 'scientific' voyage. Scientific and military interests coexisted smoothly enough in Egypt and a collaborative and fraternal union between Baudin's naval staff and naturalists was expected to lead to a further French coup for the progress of human knowledge. On the eve of the expedition's departure from Le Havre, reported *Le Moniteur Universel*,

> winds holding the *Géographe* and the *Naturaliste* in our port ... gave the savants and the mariners time to get to know each other, and friends of science saw with deep satisfaction that the most perfect union reigned between them.[38]

Indeed, the expeditioners themselves clearly had collaboration and friendship in mind. At a celebratory dinner just days before their departure, Hamelin made a toast 'to the union between the naturalists and the sailors of the expedition!', followed by Lieutenant Pierre Milius: 'may the savants of the expedition find in us their collaborators!'[39] For his own part, mineralogist Louis Depuch composed a poem:

> Long and perilous labours,
> Will soon be our lot,
> To attempt them, let us brave the seas,
> Brave the gale and the storm;
> Above all let us stay always united,
> And since a friend brings us together,
> In his honour, my dear friends,
> Let us laugh, sing, drink together.[40]

But there existed tensions between these ideals even then – tensions which were only to intensify after the expedition set sail. At Tenerife, frustrated savants and senior officers pressed Baudin for greater privileges and liberties; at Mauritius expeditioners sought escape from the hierarchy, discipline and duties their captain enforced aboard the ship; and at Timor, Baudin only narrowly averted a mutiny when his wilful first lieutenant challenged him to a duel. In Australian waters, naval and scientific interests fell more closely into step and, indeed, a union of sorts seemed to reign between the men during their sojourn in Tasmania. Nevertheless, the expedition had still not hit its stride by the end of its first campaign.

On the south coast of Australia, 1802, Baudin brought the expedition to a turning point. The *Naturaliste*, captained by Emmanuel Hamelin, had become separated from the *Géographe* for the second time during the campaign and Baudin had lost a dinghy carrying a crew of eight men, including one of his two geographers. Hamelin's commitment to the expedition's scientific mission was weak, and weakening. As he crossed Bass Strait, he contemplated a return to France, after replenishing supplies at the nearby British colony at Port Jackson. Baudin, too, considered setting sail for Port Jackson – but it was not a homeward voyage that he had in mind. Winter was setting in, and high seas and squalls prevented the *Géographe* from nearing the coastline – his charts were imprecise and his naturalists had not been ashore. Port Jackson was not a scheduled port of call and the Republic was at that point at war with Britain. However, Baudin and Hamelin carried passports signed by the British Admiralty and, during brief encounters in these waters, the haven had been recommended to Baudin by an English skipper and later by fellow-explorer Matthew Flinders. Determined to complete his instructions, the commander pushed on until the *Géographe*'s supplies were depleted, few men remained able to man the ship, and increasingly wild weather put them all in grave danger. On the morning of 8 May, he reached the point where stopped the chart of his predecessor, Bruni d'Entrecasteaux; the sky was 'so dark and overcast that one could scarcely see to work or to read in one's cabin'.[41] Fierce squalls tossed the seas beneath him that night as Baudin sheltered in the great cabin and scratched a new entry in his journal:

> Serious reflections upon the position I was in, the weakness of my crew ... our pressing need for firewood, the shortness of the days and a host of other private considerations all decided me to abandon the coast and make first for d'Entrecasteaux Channel, where the anchorage is good, and from there proceed to Port Jackson, which I have always hoped that the dinghy believed lost may have been able to reach.[42]

A sojourn in the nearby colony would indeed satisfy the expedition's immediate needs and enable it to continue on to the next stage of its voyage; however, it also offered another, more significant, advantage. The next morning, Baudin explained further:

> The three squalls that I had already endured in the space of one decade had consistently thwarted my desire to examine fully the St. Francis and St. Peter islands, which, to judge from what we have seen of them, are scarcely worth the trouble that we took over them. However, I am convinced that by exploring them thoroughly one must find some shelter for navigators who may be in this area. But to carry out this work with an appearance of success, one simply must be there in the middle of summer when the days are long and the weather is moderate. And so this is what I plan to do in the next season. The southern portion of big Kangaroo Island may also be extremely interesting, and although Mr. Flinders told me that he spent six weeks on it, it looks to me ... that he did not explore it completely.[43]

It was the exploration of Australia's south coast that most deeply interested the scientific and political institutions of France, and Baudin intended to carry it out as thoroughly as possible – even if that meant doing it twice. As he later explained to the Minister of Marine,[44] his plan was to carry out a second, more accurate, exploration of the coastline in the summer, and this plan necessitated a lengthy sojourn at Port Jackson.

This was a significant decision. Baudin chose to diverge from the itinerary – something Fleurieu had ordered him to avoid – as well as to break from the commission's desire that he visit only 'unknown' or 'little-known' areas. Moreover, soon after arriving at Port Jackson, he would disassemble the expedition prepared for him by the Institut National and rebuild it anew. As a result, Baudin's expedition was not only the first French expedition aimed specifically at Australian discovery, but also the first to sail through Sydney Heads and visit the established British colony at Port Jackson. Furthermore, the reconstructive process it underwent during this five-month sojourn – at the hands of a commander determined to satisfy the objectives of his superiors – would reveal a novel glimpse of French aspirations in the global theatre and new insight into the mechanics and culture of the Napoleonic era's voyage of scientific discovery.

2 'I SHOULD WISH ... TO ESTABLISH A FEW TENTS ON SHORE': THE PORT JACKSON STAY

While Baudin had been pushing into winter on the south coast, his second captain was turning in to Port Jackson – the British flag raised on the foremast of the *Naturaliste*. In hurried journal entries Emmanuel Hamelin and his officers recorded their arrival, noting the heavy rain and rough seas that signalled an incoming storm and that even upturned their dinghy and its crew in the harbour. The unfortunate men were left ashore for days, where they were cared for by a group of unnamed Aborigines, but the voyagers' thoughts were less on their stranded companions or those men's rescuers than they were on their British hosts. Hamelin's men had caught the distant sound of music and a nine-canon salute:[1] the colonists in Sydney Town were celebrating St George's Day and anticipating the arrival of the French voyagers – the pilot, as he guided the *Naturaliste* into the harbour, told Hamelin they had been hoping to see the expedition for the past year.[2] Still, both voyagers and colonists were unaware that the Revolutionary Wars had been brought to an end by the Treaty of Amiens and so the *Naturaliste* was anchored, for the time being, just a short distance into the harbour at rugged Middle Head. It was from there that the first official contact between the expedition and the colony was made, as, with a cautious hand, Hamelin penned his first letter to Governor Philip Gidley King.[3]

The *Naturaliste* would remain at Port Jackson three weeks before again setting sail, ostensibly, to rejoin the *Géographe*. But it was on its return to the British colony another three weeks later, having run short of provisions, that the expedition was reunited and under Baudin's command awaited summer. The scientific staff and their collections escaped the confines of each ship and spilled ashore, the naval staff adjusted to new routines and relationships, the commander turned to colonial politics and preparations for a new campaign. These were days not simply of respite and replenishment; they were busy with human encounters, conflicts and resolutions, scientific collections, observations and collaboration, personal reflections, plans and ambitions. While, examined from particular angles, their details reveal the cultural, scientific and imperial significances of this episode, they need first of all simply to be reconstructed. The everyday history of the sojourn establishes what it was like for the French voyagers as guests in the British colony. And, in particular, it gives initial insight into how Baudin and Hamelin felt about their mission in Australia and how the governor felt about accommodating a French expedition on His Majesty's territory.

The First Sojourn of the *Naturaliste*, 25 April–18 May

With his letter of introduction complete, Hamelin followed custom by sending officers – in this case, first lieutenant Pierre Milius and midshipman Desiré Breton - to inform the governor in person of the expedition's arrival and objectives.[4] King welcomed them with enthusiasm. At dinner, he declared that he saw the men of the *Naturaliste* as 'citizens of the world' with 'the right to expect the recognition of all nations', and that, although the colony was lacking essential provisions, he would take great pleasure in providing for the Frenchmen.[5] Still, the governor was conscious that this situation called for some caution. While waiting to meet Hamelin in person, he compiled a set of regulations for the Frenchmen to follow based on those applying to British and foreign merchant vessels, but more restrictive; for example, while it was only the sailors from merchant vessels who were not allowed into the countryside without a pass, all men from the *Naturaliste* – crew members, scientists and officers – needed the governor's permission to explore beyond Sydney Town.[6] No doubt King welcomed the social stimulation offered by his French guests and wanted to establish an amiable relationship with them but, with characteristic shrewdness, he was also attempting both to abide by the politics of the commonwealth of learning and to protect Britain's colonial project.

When King and Hamelin finally met in person, it was the Australian discoveries claimed by their respective nations that they were most eager to discuss. King explained, no doubt both in the spirit of cooperation and to advise the French voyagers of where they stood, that many British ships had explored this part of the world and particularly valuable discoveries had been made in Bass Strait.[7] For his part, Hamlin gave King a detailed account of the surveying work the expedition had achieved thus far.[8] Shortly after the *Naturaliste*'s departure, however, King would relay to the Duke of Portland Hamelin's account of the voyage and comment that 'the remainder of the *Naturaliste*'s voyage is a secret'.[9] Moreover, claiming that the French intended to establish a settlement on the south coast of New South Wales or the west coast of Australia, he also redoubled his efforts to obtain from the British government the means to form a colony at Port Phillip.[10] Clearly, while the relationship between King and Hamelin was congenial enough, there was a distinct undercurrent of distrust.

Politics, however, did not get in the way of festivities. Following the initial gatherings in Sydney, the governor, greeted proudly with three cries of 'Vive la

République!', was hosted aboard the *Naturaliste*, before a dinner attended by forty-eight leading members of the colony, followed by a ball and, finally, a supper at Government House.[11] Over that week, Hamelin and his officers were to dine once more at Government House and also at the home of the commissary, John Palmer.[12] It was certainly a festive beginning to the sojourn, delineated by rules of etiquette and assertions of national pride.

Neither did underlying suspicions prevent Hamelin and his men from turning to the work at hand as soon as possible. On 29 April they anchored the *Naturaliste* in Neutral Bay, opposite Sydney Cove.[13] Ill and injured men were settled into the hospital in Sydney,[14] Hamelin moved into lodgings in Sydney,[15] and the observatory tents were pitched at Green Point – far from Sydney Town, just around a mile inside the entrance to Port Jackson. Thereafter, boats daily headed back and forth across the harbour to collect bread and water, wash laundry and catch fish. Hamelin, for his part, was preoccupied with onshore tasks, such as attempting to obtain campaign supplies.

Natural history collecting was not a high priority for Hamelin during this time. He received a donation – three pieces of red cedar,[16] but the records suggest he neither purchased nor collected any objects; indeed, natural history was not Hamelin's strong point: 'it is foreign to me ... I do not claim to be clever or play the savant',[17] he had commented earlier in the expedition. However, he did once send a boat party, including botanist Théodore Leschenault de la Tour and two soldiers allocated by the colony's lieutenant governor, Lieutenant-Colonel William Paterson, to collect shells, plants and seeds. Leschenault also ventured into the countryside in the company of British botanist Robert Brown and gardener Peter Good, both just arrived in port with the Flinders expedition. The only other naturalist with Hamelin at this stage was mineralogist Charles Bailly. He carried out an analysis of ferruginous stones for Governor King.[18]

In any case, two weeks into the sojourn, as Hamelin began to despair of obtaining adequate campaign supplies for his voyage, a series of events began to take place that would seem to have augured well for he and his men. The Flinders expedition arrived in port with news of its encounter with the *Géographe* on the south coast and, most importantly, of the fact that Baudin was on his way to Port Jackson.[19] To this, was added further happy news the following day: France and England were at peace thanks to the Treaty of Amiens. King told Hamelin that, 'now, if the *Géographe* arrived at Port Jackson, it could like [the *Naturaliste*] and at the same time moor in Sydney Cove, which he did not believe he could have permitted previously'.[20]

Yet, these events seem only to have hastened Hamelin's departure. Two days following the arrival of the *Investigator*, he wrote to the Minister of Marine to inform him that, being unable to obtain the supplies he required, he intended to sail from Port Jackson the next day. He hoped to rejoin Baudin at sea and inform him of the colony's lack of resources, but, if that endeavour proved futile, then he would head directly to Mauritius.[21] This was an odd decision indeed: Hamelin knew that two merchant ships were due to arrive in the colony with ample supplies[22] and also that it was dangerous to sail south in winter. He must have known too, as Baudin himself later observed, that he had little to no chance of encountering the *Géographe* at sea.[23] In fact, it seems he did not genuinely hope to meet Baudin but intended, rather, to return to Mauritius – as Michel Jangoux opines, Hamelin had lost his enthusiasm for the expedition.[24]

However, if he was quick to give up on the expedition's mission of geographic discovery and natural history collection, conversely, he was willing to overshoot his instructions by gathering military intelligence for the French government. Before leaving the British colony, Hamelin compiled a 'Notice sur le Port Jackson', full of details that would be useful to his superiors should they consider invading the colony: how to enter the port safely, details of the population of the colony and the size of its army, the traffic of ships in and out of Port Jackson and the state of the port's defences. One copy of the report is included in his journal,[25] and another is listed amongst the various papers and charts returned to France.[26] Whether or not either copy reached its target is unknown but, obviously, Hamelin intended these observations to be laid before the powers of the French government – further motivation, perhaps, to hurry back to Mauritius. The *Naturaliste* sailed out of Port Jackson on 18 May 1802. It did not rejoin Baudin and fell dangerously low on supplies. Hamelin was forced to return to Port Jackson twenty-one days after his departure.

The Arrival of the *Géographe*

If Hamelin's conduct at Port Jackson had seemed strange, even suspicious, it does not appear to have greatly concerned King. One month after he had fare-welled its consort King readily welcomed the *Géographe* into Port Jackson. The hospitality King showed to Baudin and his men has been recounted many times and typically not without a little exaggeration. Susan Hunt and Paul Carter, for example, declare that 'King received [Baudin] and his men, nursing, feeding and entertaining them in the next five months with an extraordinary generosity'.[27] Of course, it is true that the governor did offer vital assistance to the expedition; yet, a close look at the early communications between Baudin and King reveals that the commander also deserves some credit for this *entente cordiale*.

Contact was made between the commander and the governor immediately after the *Géographe* anchored in Port Jackson on 20 June 1802. King welcomed his guests into port with a letter enclosing four documents: the proclamation announcing that peace had been established between 'His Britannick Majesty and the French Republic', a letter each from Milius and Hamelin, and the set of regulations to be followed by the Frenchmen during their stay in Port Jackson.[28] That evening, Baudin sent his engineer, François-Michel Ronsard, ashore to meet the governor on his behalf and Ronsard was welcomed just as graciously as Milius and Breton had been the previous April. It was most likely that night at dinner that King received his first letter from Baudin. Unfortunately the document appears to have been lost, but it, or perhaps just the joyous mood of the moment, inspired on the part of King an assertion of generosity contradicting his earlier statement to Hamelin. On 21 June, King declared:

> I had the honour of receiving yours of yesterday's date, and altho' last night I had the pleasure of announcing that a peace had taken place between our respective countries, yet a continuance of the war would have made no difference in my reception of your ship, and affording every relief and assistance in my power.[29]

That morning, with some assistance from a group of sailors provided by the port, the *Géographe* reached its anchorage in Neutral Bay.[30] Early that evening, Baudin went ashore,[31] where he undoubtedly familiarized himself with the resources available in the colony, then visited Flinders aboard the *Investigator*.[32] The next morning he laid out his requirements in a letter to King.

This approach clearly differs from that taken by Hamelin, who made his requests before stepping ashore and, indeed, before any other direct contact with the governor had been made. The letters, wherein the two French captains made their initial requests, are also dissimilar. While Hamelin's first letter to King has a diffident style, the letter written by Baudin on 22 June conveys a sense that the commander knew exactly what he was doing and what he wanted. In describing the poor health of his crew and requesting that they receive treatment in the colony's military hospitals, Baudin appealed to King's sense of humanity and also acknowledged the governor's experience in such matters, for example by commenting: 'this disease, as you know, requires only some nursing, tranquillity, and a dietetic change'. His following requests were phrased more directly and indicate the type of information he must have been gathering on shore or from Flinders:

> I should also wish, subject to your approval, to establish a few tents on shore to facilitate the work of our astronomers, whose observations shall be communicated to you. The place where Mr Flinders is located appears to be the most convenient provided you see no objection.

It is interesting to note, here, Baudin's diplomatic astuteness in offering to share with King the astronomers' observations. Continuing with his confident approach, Baudin ended:

> As I shall be compelled to take some provisions such as biscuits, flour, salt meat, spirituous liquors, fresh meat, vegetables etc. etc. I shall have the honour of forwarding

you the list of quantities, praying that they should be supplied to me from the Government or private stores, if there exist any.[33]

King responded, granting Baudin's requests, in equally direct style.[34] The frank tone of the letters between Baudin and King reflects something of their respective characters, but also sets the tone for their relationship and the sojourn as a whole.

Before heading ashore again for the day and making a second visit to Flinders,[35] Baudin handed Ronsard the regulations compiled by King as well as a list of complementary orders that he had put together himself. The very precise nature of Baudin's orders concerning his men's movements and conduct on the harbour and in town would seem to reflect local advice. He stipulates, for example, that:

> no boats on fishing expeditions may pass to the west of the ship but only to the east, we may fish in any of the harbours to the north or south, in the last case it is expressly prohibited to go ashore without written and signed permission from the Governor, who will designate where it is permitted to go ashore.[36]

Baudin ordered that anyone who wanted to move into lodgings ashore needed to put their request in writing to King and, once ashore, if by their 'conduct or indiscreet curiosity' they received a reprimand, they would be required to return to the ship.[37]

It is often noted that, despite warmly receiving the French expedition, King was acutely conscious of the caution required in this situation;[38] clearly, Baudin was equally cautious. The Minister of Marine and the Colonies had instructed him on how to behave in foreign ports[39] and he was evidently aware of the strict protocol that must be followed, particularly in a port that belonged to France's greatest rival and that was located in a region of considerable strategic importance to that nation. The following day, 23 June, Baudin wrote to King offering to lay before him all the paperwork relating to his voyage.[40] His subsequent visit with the governor and the lieutenant governor marked the end of the initial and most important formal negotiations between King and Baudin;[41] subsequently, their association became considerably more relaxed.

It was also after this date that Baudin and some of his men installed themselves on shore.[42] On 25 June, tents were erected on Bennelong Point – one for the sailmakers and two for the observatory – and the astronomer, Pierre-François Bernier, set up the astronomical instruments. The next day, Bernier established himself at the observatory where he would stay for the duration of the sojourn and where, at Baudin's orders, the officers of the *Géographe* would ensure that he received his meals every day.[43] Baudin himself rented a house in Sydney,[44] and sailors from the *Géographe* brought daily supplies of bread and vegetables to the governor's dock for him.[45] The naturalists too settled into lodgings in town. At the end of this first week the crew began to disembark the natural history collection: initially, the plants were taken ashore, some to be placed in the sailmakers' tent and others at Baudin's residence, and later the zoological crates were also unloaded.[46]

One of Baudin's highest priorities upon reaching Port Jackson was to arrange medical care for his men. Typically, the treatment of the French patients during this sojourn is represented within the context of British hospitality. However, the voyagers did not improve 'with astounding speed [simply] as soon as they returned to a rich diet of fresh produce'.[47] Their recovery was due to the combined effort of the colony's principal surgeon, James Thomson, as well as Baudin himself and the French doctors, Hubert Taillefer and François-Etienne Lharidon.[48]

As soon as the ill and injured men were settled into the Sydney hospital, Baudin ordered some of his sailors to supply them with hammocks,[49] set up a bath for them on shore[50] and deliver their rations:[51] one pound of bread, half a pound of meat and a quart of wine.[52] This food was provided in addition to the regular diet instituted by Thomson, which consisted mainly of vegetables and included daily servings of lemonade,[53] and which was paid for by the commander: £20 for refreshments, £15 for vegetables.[54] As customary during a sojourn in a port of call, Baudin also played a role in obtaining medicines for his men – he purchased bark, 'glauber salts',[55] epsom salts, lint,[56] glyster pepis[57] and seventy-two bottles of lime juice to a total of £76[58] – but he made a particular effort during this sojourn: a letter addressed to Baudin from Matthew Flinders indicates that Baudin had been seeking a medicine named 'cincona'[59] and had hoped that the surgeon travelling with Flinders could provide him with some.[60]

Moreover, Baudin required that Taillefer and Lharidon keep him well-informed of the patients' progress. They thus produced comprehensive documentation – inventories of medicines purchased, tables recording the daily movements of patients in and out of hospital, medical reports and death certificates – and Taillefer compiled daily reports, commenting upon such matters as individuals' level of pain, the fading of their sores and the severity of their fevers.[61] One of the most interesting documents is a letter from Lharidon to Baudin, written two weeks after their arrival in port. The patients had not been progressing as well as they ought to have been, declared Lharidon: those most in need of care had complained of not receiving the attention they required while 'the others appear to be leading a life that will not allow them to enjoy' their improved health for long. Lharidon believed that he would be able to treat the patients with greater success if two-thirds of them were returned to the ship, leaving in hospital only those who were genuinely bedridden: 'without these precautions', warned Lharidon, 'debauchery and greed will convert the mildest conditions to interminable diseases'.[62]

Despite Lharidon's prediction, the patients' conditions did not generally worsen; in fact, Taillefer's reports indicate that their health overall was improving. If Baudin had concerns about the situation, it would most likely have related to the apparent inappropriate conduct of the men who were supposed to be in hospital. In any case, on 24 August the thirteen patients remaining in the Sydney hospital returned to the *Géographe*.[63] Thereafter, it appears that members of the expedition were treated exclusively by the French doctors in a tent erected and set up as a hospital on shore.[64]

The Expedition Reunited

The *Naturaliste* made its second arrival at Port Jackson only days after Baudin had placed his ailing men in the Sydney hospital and he and the scientific staff had settled into their lodgings. Of course, Baudin knew all about Hamelin's earlier visit to the colony, but his most direct source of information was Milius, who due to ill health had stayed behind after the *Naturaliste*'s precipitous departure. He had paid a visit to Baudin on 23 June and explained to him why Hamelin had decided to leave the colony instead of waiting for the *Géographe*. Recounting the meeting afterward, Milius wrote that none of the reasons he gave seemed to please Baudin very much.[65]

Indeed, it is not difficult to imagine just how displeased the commander must have felt; however, if he was angry with Hamelin, his papers only hint at it. He was critical, yet brief, when he wrote of it to Jussieu: 'I regret the departure of the *Naturaliste*', he wrote, 'which could not have met us but by a very unlikely chance given the season in which we found ourselves'.[66] In another letter to Jussieu, in which Baudin pointed out that his second captain had actually intended to reach Mauritius, he refrained from expressing an opinion.[67] As a number of commentators suggest, while Baudin may have been perplexed or even taken aback by Hamelin's decision, such feelings do not appear to have affected their relationship.[68] However, Baudin did soon decide to exclude Hamelin from the second campaign and, instead, send him back to France with the responsibility of delivering, aboard the *Naturaliste,* the natural history collection from the first campaign. If, indeed, Hamelin had long been eager to 'escape a dependence on Baudin that injured his pride', as Saint-Cricq claimed,[69] perhaps Baudin respectfully recognized his unsuitability for the expedition and decided to allow him to return with dignity.[70]

It has been proposed that Baudin's reunion with Hamelin seems to have restored some of his energy.[71] He certainly must been relieved to have the *Naturaliste* and its men with him again, and to be assured of their well-being. Of course, the return of the *Naturaliste* happened also to be timely. At this stage, with the entire expedition at hand, he was in a position to plan and prepare for the new voyage. Thus, expedition turned to the work at hand and settled into daily life in Port Jackson.

The Naval Staff at Work and Leisure

The sailors' days during the sojourn were, naturally, different from those at sea, but they were nevertheless busy and closely supervised. Under the immediate direction of the officers, who, in turn, reported to Baudin, the sailors tended to regular daily chores as well as preparations for the next voyage.[72] Of their experiences, more precisely, we have only the journal and logbook records of their superiors, which mention the sailors only in certain circumstances: generally, when their actions influenced conditions or relationships aboard the ship and, typically, resulted in punishment. Consequently, it is difficult to develop an accurate picture of their daily life in Port Jackson and, indeed, this is why they tend often to be left out of the narrative.

An impression of the sailors' various responsibilities and a hint of their lifestyle may be gained by looking at the list of items that Baudin purchased at Port Jackson for the boatswains. It includes mainly work-related items, such as nails, fishing line and hooks, tar, brushes, black and white paints and a wooden frame for holding plants, and only a few personal products, such as soap and candles. As noted earlier, the sailors were responsible for delivering provisions to Baudin, to the patients in hospital in Sydney and also to the astronomer and the guard watching over the tents on Bennelong Point. In addition to these tasks, a group of sailors was required to row to shore each day to collect provisions for the men aboard the ship, others regularly collected water and gathered firewood, while every two days sailors brought aboard the ships' rations of fresh bread.[73] For certain tasks, such as refitting, repairing and stowing supplies onto each ship, teams of sailors from the *Naturaliste* and the *Géographe* worked together.[74]

After working for over a month, from July to August, to replace the copper sheathing on the *Géographe*, it was time for the sailors to receive their pay. However, Ronsard and Hamelin, in the absence of Baudin who had left for Parramatta, felt it would be wise to delay payment until all of the equipment that had been placed ashore while the *Géographe* was being repaired had been loaded back on board. 'It would have been impossible to get anything done once the sailors had been paid', remarked Ronsard. He gave them their money the following morning and 'that evening the whole crew was drunk'. At 9.30 pm, after returning from Sydney in a drunken state, petty officer Pierre Ector was killed by a fall from the gangway on to the gun-deck.[75]

Nevertheless, the sailors were given many opportunities for recreation. Frequently, groups of around ten were sent ashore to enjoy a day in town and they certainly took advantage of this privilege. The officers aboard the *Géographe* frequently reported that certain sailors had not returned to the ship before the curfew, which was prescribed by King in the regulations he gave to each ship and reinforced by Baudin's own orders. Moreover, there were frequent incidents of misconduct – usually related to drunkenness – while they were on shore and, to a lesser extent, aboard the ships. Baudin's style of discipline was fairly consistent and rarely violent. In accordance with his direction, the officer on duty typically disciplined sailors by either confining them to the ship or locking them in irons for a length of time.[76]

However, some crimes required more severe punishment, and the way in which Baudin dealt with these gives us some indication of his eagerness to maintain discipline aboard his ships and to comply with the regulations of the colony. The incident that provides the most fitting example occurred less than a month into the sojourn. On the night of 15 July, Baudin summoned the officer on duty, Henri Freycinet, to the governor's dock. He informed Freycinet that several pieces of sailcloth and some gunpowder from the French ships had been sold on shore and ordered him to conduct a thorough search throughout the *Géographe*, excluding the locked chests, in order to find any further items that may have been hidden. Freycinet carried out the order immediately but the search was unsuccessful.[77] The next day, the Sydney police arrested the suspected thieves on shore: master gunner Valentin Kleinne and gunners Yves Menou and Vinard Barbier from the *Géographe* and master gunner François David from the *Naturaliste*.[78] Baudin consequently ordered another search, which, this time, was to include the chests belonging to Kleinne. With the assistance of the sub-commissary, the first and second boatswains and two master gunners, the officer on duty found two pieces of sailcloth and 35 pounds of gunpowder.[79] With permission from King, and in accordance with British naval ordinances, Baudin then made arrangements for a court martial to be held aboard the *Géographe* on 20 July.[80] Hamelin and midshipman François Heirisson came aboard to form the jury, soon followed by the harbourmaster, John Harris, with the witnesses and the accused. Ronsard acted as the president of the jury and declared the unanimous verdict, namely, that Kleinne and Barbier were guilty of stealing four pieces of cloth and that David and Menou were not guilty. For their crime, Kleinne and Barbier were placed in the Sydney prison where they stayed for the remainder of the sojourn.[81]

The officers, in charge of the crewmen and responsible for all the work to be carried out aboard the ships, were supervised by Baudin with particular care. Their days were regimented and they were constantly held to account for their own actions and for those of the men working with them. The logbook that was kept aboard the *Géographe* throughout this episode provides valuable insights into the daily lives of these men. Early in the sojourn, Baudin gave the lieutenants, Henri Freycinet and Ronsard, and the sub-lieutenants, Gaspard Bonnefoy and Joseph Ransonnet, individual responsibilities concerning the maintenance and movements of the ship. They also took turns acting as the officer on duty aboard the *Géographe*, and at the end of each day, in this role, they submitted to Baudin a report, entered into the ship's logbook, concerning the day's events.

They recorded therein the departures and arrivals of each boat party and their purposes: taking men to Sydney or returning them, fetching or delivering provisions or cases of natural history objects, or visiting another ship such as the *Naturaliste* or one of the foreign merchant vessels. They commented on the work being carried out that day, such as moving the ship's fittings to the tent on shore or careening the *Géographe*. Finally, they included in their reports notes concerning orders, reminders and visits from Baudin. Baudin required the officers on duty to send these reports to him daily, but there are entries in the log and in Ronsard's journal showing that they often submitted them in person. This gave the officers an opportunity to discuss with their commander any particular issues that had arisen that day.[82]

Among the aforementioned regulations that Baudin set for his men at the beginning of the sojourn was one requiring that there always be two officers present on the ship; yet, subject to a curfew and on the condition that they wore their uniform,[83] they frequently rowed ashore to spend the day at leisure in Sydney Town. After six long and laborious months at sea,[84] the French officers and savants took advantage of these opportunities to indulge their gentlemanly sensibilities with European luxuries. The accounts of expenses accrued during the sojourn show that Baudin permitted the officers and scientific staff to spend certain amounts of money, upon request, and that most of their purchases were made at the warehouse of Sydney merchant Simeon Lord.[85] Baudin's account with Lord lists a vast variety of items, from utilitarian objects bought in large quantities, such as fish hooks, to personal products purchased in smaller quantities, probably by the officers, scientists and artists, such as 'fine' hats, green tea, combs and stockings. These types of purchases were numerous: over one month, a total of ninety-three silk handkerchiefs and sixty-seven yards of hair ribbon were added to Baudin's account.

Unfortunately, the records kept by Ronsard, Henri Freycinet, Bonnefoy and Ransonnet provide us with little insight into their social lives. We do know, however, that as the *Géographe* was moored alongside the *Investigator* and the tents from these two ships were together on Bennelong Point, there was some interaction between the officers of the Baudin expedition and those of the Flinders expedition. John Franklin, midshipman aboard the *Investigator*, wrote to his sister that 'we were in company at Port Jackson with the French Discovery ship, and, "unfortunate me", was obliged to converse with their officers in Latin'.[86] The two expeditions were together at Port Jackson for one month, during which time, if language difficulties did not prohibit their communication too much, the officers would have had the opportunity to compare notes on their work and on their expeditions more broadly.[87]

It was, however, their contact with the officers of the New South Wales Corps that impacted most significantly on their experience at Port Jackson and even impacted on colonial politics. Baudin's officers were in regular contact with the British marines, at formal dinners, no doubt on the streets of Sydney and perhaps too at local meeting places such as the 'sort of café ... where there [were] various games – notably billiards'.[88] Over these five months, a degree of familiarity thus developed between the two groups. Ronsard claimed a friendship with lieutenant George Bellasis; though, on this point it is perhaps significant that Bellasis was something of an outsider in the colony. Formally an artillery officer for the East India Company in Bombay, he had been transported to Port Jackson after he killed a man in a duel, in defence of a woman's honour. King had emancipated him and appointed him as a lieutenant of artillery but, largely because he was an ex-convict, he was not enrolled in the New South Wales Corp. When the French ships arrived, Bellasis had been in the colony only six months. It is unclear how friendly the Frenchmen became with official and more established members of the Corps. However, there is evidence of suspicion and resentment harboured amongst the British officers. Lieutenant James Hingston Tuckey, who would later take part in the abortive attempt to establish a British settlement at Port Phillip Bay, mentioned the French expedition in the context of his concerns about the colony's lack of defences:

> That this remote colony has not escaped the thoughts of our enemy is clearly demonstrated by two French frigates the Geographe and Naturaliste remaining at Port Jackson two or three months, during which the officers minutely surveyed the harbour and also made scientific excursions into the interior by which they acquired more extensive knowledge of the nature of the country than the colonists themselves.[89]

Furthermore, rumours about the Frenchmen and about the true objectives of their expedition were rife in Sydney Town during these months. When colonial authorities and Baudin came into conflict over a difference in the naval protocols of their respective nations, tensions were fuelled by local gossip concerning the honour of the French officers. And later, a member of the New South Wales Corps drew Baudin's officers directly into colonial affairs when he accused certain among them of trading in spirits and thus contravening a regulation highly unpopular with the Corps. Despite the continuing demands of life aboard their ships during this sojourn, in fact, the French officers became quite immersed in the world of the colony.

Voyager-Naturalists in the Colonial Field

A considerable amount of Baudin's time was clearly taken up by supervision of his naval staff, and undoubtedly also by preparations for a return to sea, but he would also have interacted frequently with the savants. They were all lodging in Sydney, they attended social gatherings and Baudin, both as their commander and as an enthusiastic natural history collector, was deeply interested in their work. It is uncertain whether Baudin gave the scientific staff any close direction during this period but there is evidence that he did provide them with materials and opportunities to pursue their research.

During their time in Port Jackson, each of the naturalists – botanist Leschenault (presumably with the gardener Antoine Guichenot), mineralogists Bailly and Louis Depuch and zoologist François Péron – took opportunities to venture beyond Sydney Town. They explored Botany Bay, Parramatta, Castle Hill, Hawkesbury and the fields, forests and rivers in between, up to the foot of the Blue Mountains. To facilitate the Frenchmen's research, colonists provided transport, guides and accommodation for these excursions. However, it is not clear whether or not they did so at their own expense. Baudin's list of expenses accrued during these months lists payments made for a voyage to Hawkesbury (£10), another to Parramatta (£3) and one to the Blue Mountains (£10),[90] but these are places that Baudin visited himself on an excursion with the governor in August and, therefore, may well relate to his own expenses. If the naturalists' excursions did incur any costs, then, they may have been paid upon request and been included in the expenses accrued by each individual during the sojourn.[91] However these excursions were financed, and even though British naturalists had been studying the environment of the County of Cumberland since the arrival of the First Fleet in 1788, they allowed the French naturalists to amass vast collections of specimens.[92] Additional equipment was in fact required to hold and preserve them: bird cages, glass cases, pots and 'garden stuff'.[93]

The work produced at Port Jackson by the artists Charles-Alexandre Lesueur and Nicolas Martin Petit was equally significant. The months on shore allowed them to take more time and care with their drawings than they were usually afforded by fleeting beach excursions and the cramped, unsteady conditions of the ship. Of course, the colony also presented them with scenes they would not witness anywhere else during the expedition. For example, Lesueur produced a series of sketches of Sydney and of Aboriginal artefacts particular to the Port Jackson region, while Petit drew a number of detailed portraits of local Aboriginal people.[94] Whether or not Baudin gave the artists any direction is almost impossible to gauge, yet, even if he did not, it is likely that he at least had some knowledge of the type of sketches they were producing and tacitly approved.

The geographers Pierre Faure and Charles-Pierre Boullanger and the astronomer Pierre-François Bernier were very quiet during this period. Neither Faure nor Boullanger made any journal entries during their stay at Port Jackson and we do not know whether Baudin assigned them any responsibilities.[95] However, Boullanger found an opportunity to utilize his skills by updating an existing plan of the town of Sydney.[96] Bernier, for his part, worked assiduously throughout the sojourn – his journal entries relate almost entirely to his astronomical observations and his work was valued very highly by the commander. Baudin granted Bernier's request for the expedition's departure to be delayed two weeks – a long time when maritime 'discovery' awaits – until he had had the opportunity to observe the transit of Mercury from his current position rather than,[97] as had been planned, during the later survey of the south coast.[98]

Hamelin's Second Sydney Sojourn

For the scientific staff, the Port Jackson sojourn represented a respite from life at sea, with access to the comforts of a European settlement and opportunities to pursue more sustained and leisurely research than typical voyage conditions permitted. By contrast, for Hamelin it was a chance to focus on shipboard tasks. It is important to remember certain points about Hamelin and Port Jackson: this was his second visit to the colony, he had been rather unsettled during the first sojourn and, eager to return to Mauritius; he had returned to Port Jackson with considerable reluctance. Moreover, although it was some time until Hamelin was informed of it, he was to be sent by Baudin directly back to France. His approach to this sojourn is therefore of considerable interest, all the more so as it has been somewhat overlooked in accounts of the Port Jackson episode.

Hamelin became preoccupied with life aboard the *Naturaliste* during these months: almost one month after arriving in Port Jackson, he was still living aboard the ship, and on 22 July he noted that Baudin had ordered him 'to spend night and day on shore to assist him in his work, consequently I had my bed brought ashore and I went aboard each day'.[99] The remainder of his journal entries relate entirely to business aboard the *Naturaliste*. And although, where appropriate, he notes the orders received from Baudin, he makes scant mention of his interaction with the commander and his activities ashore.

It is tempting to surmise that Hamelin kept mainly to the ship in a desire to avoid Baudin's command and also to avoid the British colonists, whom he seemed to distrust during his previous visit. However, there is no evidence that his relationship with Baudin or the colonists was unfriendly. Rather, the relaxed tone of his journal entries suggests that he was more at ease during this second sojourn, in his shipboard role and under the command of Baudin, than he had been during his first visit to Port Jackson. Although in some ways it appeared that, prior to this sojourn, he had wished to escape from his subordinate position, it is also true that, in explaining to his commander his decisions to sail for Port Jackson and then leave for Mauritius, he seemed eager to be at Baudin's orders.[100] With Baudin present to manage the expedition, Hamelin had orders to follow and markedly less responsibility, which, if he had lost enthusiasm for the expedition, must have eased his mind. Baudin also provided for him, giving him £281.8*s*.2*d* to cover meal expenses.[101] Thus, Hamelin could immerse himself in the world of the ship – his preferred milieu.

Life aboard the *Naturaliste* during the first half of the sojourn was reasonably quiet. Throughout July, Hamelin's men were mainly occupied in rendering assistance to their counterparts aboard the *Géographe*, who were working on repairing the copper sheathing of the ship. On 2 August, Baudin informed Hamelin that he was intending to send the *Naturaliste* back to France with the natural history collection[102] and, upon learning this news, Hamelin set about arranging for the ship to receive new rigging; otherwise, his journal remains rather quiet over the following few weeks, containing only brief notes relating mainly to the weather. From late August until mid September, Hamelin's journal entries describe the attempts his men made, under his supervision, to rid the *Naturaliste* of its infestation of rats.[103]

By the time this task had been accomplished, as far as was possible at least, the *Casuarina*, which was to replace the *Naturaliste* in Baudin's second campaign, was afloat in the harbour and Hamelin took pleasure in helping to fit it out for her impending voyage. He recorded in his journal that he gave its captain, Louis Freycinet, all the charts and plans he had received from the French government for the expedition and allowed Freycinet to take his choice of books from the library aboard the *Naturaliste*.[104] He declared that 'it is the *Naturaliste* that supplied the *Casuarina* with her rigging, sails, jetsam, half her water, her biscuit and generally all her armament for more than a year's campaign'.[105]

Hamelin must have been kept quite busy for the remainder of the sojourn, as many of his staff members transferred to the *Casuarina* and the *Géographe* to undertake the second campaign, while some of the *Géographe*'s men boarded the *Naturaliste* to return to France. The natural history collection was packed and supplies that the *Naturaliste* would no longer require, such as oil, salt, paint and sail cloth, were passed on to the *Géographe*.[106] On 14 November, shortly before the expedition's departure from Port Jackson, Governor King, Captain Kent[107] and George Bass came aboard the *Naturaliste*. As the ship did not carry any canons, Hamelin himself saluted King with three cries of 'Vive la République'.[108] His final experience of Australia seems, thus, to have been a positive one – he was to make an honourable return to France.

The Commander in Port

As far as Hamelin was concerned, the expedition was reaching an end, but for Baudin, it was far from over. He needed to design the itinerary and determine the objectives for the second campaign, and he did not waste any time. We know that Baudin's purpose in visiting Port Jackson was to prepare for a re-examination of the south coast – that region he had highlighted himself in his initial voyage proposal and to which the Institut and the government had subsequently given distinct priority. The first attempt at this survey had been affected by inclement weather and the inability of the *Géographe* to manoeuvre close to shore, and Baudin was unsatisfied with the results. He felt that the *Naturaliste*, which was just as bulky as the *Géographe* and was also very slow, would only hinder the work he intended to undertake. Sending the *Naturaliste* back to France would also allow him to deliver the already immense natural history collection to the French authorities and, at the same time, return to France the men who were unfit or unsuitable for the remaining voyage. And purchasing a smaller vessel, able to draw nearer to the coastline than either the *Géographe* or the *Naturaliste*, would enable his men to produce far more accurate charts.[109] It was therefore very soon after his arrival at Port Jackson that Baudin set about procuring such a vessel. Finding one in the dockyards that was newly constructed and appeared to suit his purpose, he made enquiries with its owners, Sydney shipbuilder and merchant James Underwood and his partner Henry Kable. On 7 July they wrote to him with their proposals for her completion and a list of the articles they expected for the purchase.[110] Being happy, presumably, with these conditions, Baudin then sought permission from the governor to make the purchase and, on 11 July, received King's acquiescence, 'as it [was] for the advancement of science and navigation'.[111]

It was probably once these decisions had been made that Baudin began to concentrate on the itinerary for the second campaign. Upon his arrival at Port Jackson, according to the accounts he gave in his journal and in his letters to Paris, Baudin planned to return to the south coast to examine King Island ('the land that is said to exist to the north of the Hunter Islands') and the southern part of Kangaroo Island as well as to rechart the St Francis and St Peter Islands.[112] While at Port Jackson, however, his plan became more detailed and ambitious. Before leaving the colony, he explained to Jussieu and the Minister of Marine that he intended also to examine the coastline between the St Peter and St Francis Islands, return to the St Vincent and Spencer Gulfs, and re-examine Géographe Bay and the north-west coast from Shark Bay onward, finishing, finally, at the Gulf of Carpentaria.[113] Clearly, his new plan was based on a determined desire to perfect the geographical work at hand. Yet, he had not forgotten the scientific objectives of the expedition: 'I will do my best to gather a new collection as large as that which you are going to receive by the *Naturaliste*', he declared to Jussieu.[114]

Baudin's days in Port Jackson seem to have been devoted mainly to preparation for what Louis Freycinet referred to as the 'new expedition'.[115] He needed to prepare the *Géographe* and the *Casuarina*, along with their men, for at least another twelve months at sea. He also needed to get the *Naturaliste* ready to return to France with her precious load of natural history objects. As we have seen, in order to facilitate the sojourn that was now essential to his improvised plan, he closely supervised his officers and crews in their work and in their conduct as guests of the British colonial administration. He also undertook a revision of his staff and announced the results near the end of the sojourn, in early November. He allocated all of the naturalists who were fit to continue to the *Géographe* and Louis Freycinet, first lieutenant aboard the *Naturaliste*, was given the command of the *Casuarina* with Léon Brèvedent as his midshipman and fourteen 'elite' sailors from the *Naturaliste* as his crew. Fitting out the *Casuarina* and preparing Louis Freycinet for the task ahead absorbed a considerable amount of Baudin's time. There were lengthy negotiations with Freycinet concerning the equipment with which he was to be provided and ongoing discussions concerning both his new position and the role of the *Casuarina* in the second campaign.[116]

Acquiring the necessary campaign supplies for each vessel was one of Baudin's most important and, it seems, difficult tasks. He needed to build relationships with various merchants in the colony and negotiate with them concerning the quantities he could obtain and, most importantly, the method of payment to be utilized. Over the duration of the sojourn, he traded with many merchants residing in the colony: James Underwood and Henry Kable, Simeon Lord, William Cox, Samuel Enderby and Thomas Palmer, as well as some farmers such as Andrew Thompson and Samuel Skinner. He also traded with fellow mariners Richard Brooks, captain of the convict ship the *Atlas*, and George Bass, aboard the *Venus*. By the end of the sojourn, the quantity of livestock and produce that Baudin had procured for his vessels must have been immense; one purchase alone, from Skinner, consisted of twenty pigs, eighty full-grown and twenty half-grown fowls, twelve ducks, two goats and thirty-eight bushels of maize.[117] Most of the campaign provisions were purchased with letters of exchange, which came to a total of £9,374.7*s*.11.5*d*.[118] However, barter being a customary method of payment in the colony at this time, Baudin also traded quantities of rum, to a total of 122 gallons, for various objects, products and services from local individuals.[119] Moreover, he stated to King that

> far from giving the rum for its own value in the country, I have quoted it at 10*s*, so that those who have procured me specimens of natural history and provisions should get a profit which would induce them to serve us well.[120]

During the first campaign, the expedition had struck problems with supplies running short and perishing, which affected the strength of the crew and, therefore, the quality of their work. Baudin was anxious to prevent this from occurring again: before embarking on the second campaign, he wrote to Jussieu saying:

> I fear that all this work will last longer than the provisions that I have obtained here, because the geographic observations require a great deal of time; and surveys completed too quickly will be superficial, imperfect and full of errors.[121]

Despite all these responsibilities, Baudin did pause to look around and engage with his surroundings. He gathered and recorded information about the administration of the colony, as did several of his men, noting, in particular, its remarkable growth and prosperity. His observations did not have the military flavour that characterized those recorded by Hamelin and some of the other Frenchmen, nor did they focus upon the penal system: instead, they concerned Port Jackson's commercial and strategic potential. He assessed the agricultural productivity and mercantile enterprises of the colony and pointed out to the Minister of Marine that 'the colony should fix the attention of the Government and even the other powers of Europe especially Spain'.[122] In August, when the replacement of the *Géographe*'s copper sheathing had been completed, King escorted Baudin on a tour, over five days, around the colony; they journeyed up the Hawkesbury River to the township of Parramatta and to the foot of the Blue Mountains. On his travels, Baudin was given some native seeds from inhabitants of the colony and he collected some others himself, which, altogether, filled a crate.[123] In fact, while this was his only opportunity to collect specimens in the field, Baudin did procure a considerable number of natural history objects by other means during the Port Jackson stay. The expense accounts show that he purchased several live animals and plants, and through his correspondence we discover that he received some substantial donations from members of the colony.[124] Baudin also clearly admired the local landscape: 'as I write, the whole countryside is in flower, the sight and beauty of which are unparalleled; for variety, I know only the Cape of Good Hope that could be compared to it', he exclaimed to Jussieu.[125] And he took time to observe the situation of the Indigenous people of the area: 'Most have retreated far into the interior of the country to live in their own way; others usually wander through the town and the countryside'.[126]

The Aboriginal people of Port Jackson were a part of the Frenchmen's day-to-day life during these five months on shore; while this was suggested in the artwork by Lesueur and Petit, it could be easily overlooked when reading the voyagers' records. They were near Baudin and his men in town, in the hinterland and on the water, they were frequently in direct contact with their visitors, throughout the duration of the sojourn. As historians often mention, however, the Frenchmen wrote surprisingly little, or, more precisely, at surprisingly little depth, about them. They had been enthralled rather by the Europeanized aspects of the colony and absorbed by the more straightforward objects of the expedition.

The voyagers were so preoccupied and busy, in fact, that few maintained their journals throughout these five months. The commander himself gave an account of the sojourn only in his correspondence with Jussieu and the Minister of Marine. All the same, an assortment of letters, expense accounts, logbooks, journal entries, *petits mots* and colonial correspondence altogether thread a detailed and rich story of the voyagers' days at Port Jackson. They reveal the routines and hierarchy of the shipboard world largely maintained, if different than when at sea; the natural history project prioritized under Baudin's command, if not Hamelin's; and the Anglo-French relationship defined by naval rules of etiquette, questions of national and manly honour, as well as the ideals of the Republic of Learning. There is also the glimpse of an expedition in transition, as these documents detail Baudin's preparations for the 'new expedition' and sketch out the climactic tension that existed for some of his men between military and 'discovery' ambitions.

3 DISCIPLINING PASSIONS: FRENCH NAVAL-VOYAGERS AT ANCHOR

Stopovers in port always brought ambitions to the fore. Opportunities arose when the ship was moored and the voyage on hold. This was when the captain announced promotions and, sometimes, demotions; on occasion, he would order a rebellious marine to disembark and await the next ship back to Europe. The minds of his officers and sailors thus turned inevitably to thoughts of advancement or escape, authority or freedom, futures in discovery or combat. And, just as inevitably, on deck, the competing and mismatched combinations caused both collaboration and tension. Expeditions are notorious for the tense relations that typically pervaded their cramped shipboard worlds, but, in port, these tensions were considerably heightened. They were fuelled by the anticipation of staff changes, certainly, but also by changes to routine, responsibilities and the extent and use of space aboard the ship, as well as by access to the facilities and society on shore. Although the business of the expedition continued to demand attention, the busyness slowed, and the cultural assumptions and ideals of home – which, at sea, had been somewhat subordinated to naval culture – rose more distinctly to the surface. The ship in port, as Greg Dening highlights, was an ambivalent or liminal space.[1] But each port was different, each was – as Dening may have termed it[2] – a different 'theatre', and thus each port inspired different concerns, aspirations and behaviours. At Port Jackson, Governor King referred to Baudin's officers as 'citizens of the world', but they were also French naval officers visiting a British colony. This particular theatre would highlight concerns about personal ambition, patriotic concepts of French masculinity, honour and glory, as well as relations of dominance on deck and at the governor's table.

This would seem to have been a potentially fraught scenario; particularly given that, as a rule, the longer an expedition remained in port the harder it became for the captain to maintain order, that is, to maintain the shape and drive of the expedition. During a five-month sojourn in Tahiti, in 1789, Captain William Bligh lost control entirely. Yet, just as Port Jackson was not Matavai Bay, Baudin was not Bligh. Moreover, Baudin's men were of a different time – a particular ideological moment in their nation's history. Somewhere in the combination of these circumstances is the reason why, although this was the longest sojourn of the voyage, it was also – if not without its tensions – undeniably the most peaceful. In the way order was maintained aboard the *Géographe* during these five months, and in what went on below its surface, is the naval element of the turn-of-the century French expedition.

The Republican Navy

Indeed, it is important to remember the overlap here between the histories of French discovery and of the French navy. Most of Baudin's officers and sailors had stepped onto this expedition straight from service in the Revolutionary Wars. Many were serving in a natural history venture for the first time while others had sailed with Baudin on his previous botanical voyage to the Caribbean or, earlier, with Bruni d'Entrecasteaux. Moreover, with the exception of Ronsard, the engineer, the officers were no older than twenty-three and had all joined and risen through the ranks of the navy during the years of the French Revolution – a critical period for the French navy.

Not only had the concepts of equality and merit been introduced officially and comprehensively to the fleet but, from the early years of the Revolution, naval staff had no longer defended a leader but the sovereignty of the people. Moreover, like their compatriots in the army, they had come to see their roles as sailors and citizens as one and the same and it was less the man above them that they obeyed than *la patrie*. During the Revolution, the Republican fervour of seamen and political divisions among officers had caused considerable disorder aboard French ships; however, William Cormack suggests that, as the 1790s drew to a close, a return to executive power was beginning to be reflected in a more disciplined, united and purposeful fleet.[3]

This is not to suggest that, as they sailed under the authority of Bonaparte, Baudin's seamen or officers was necessarily imbued with a heightened confidence. The French navy had been weakening as a result of institutional deficiencies and, reveals Cormack, naval officers shared a pessimistic view of its history and were determined not to repeat it.[4] The crushing defeat of Bonaparte's fleet at the Battle of the Nile, in particular, must have been fresh in their minds. Furthermore, concern about naval performance was central to broader imperial anxieties, which, broadly speaking, were not unique to France but were intensified there as the nation emerged from revolution. The martial masculinity promoted during this era and its particular manifestation in naval manhood were largely expressions of the ferment of ambition and apprehension that characterized this moment of French imperialism.[5] Indeed, a number of issues were at stake aboard the *Géographe* at Port Jackson – issues which were integral to the state of the French navy,

French imperialism and this French expedition and which, during this episode, could both threaten and preserve the state of order on deck.

The Ship at Anchor

During the expedition's first days at Port Jackson, the space inhabited by the officers and crew of the *Géographe* was significantly expanded and reshaped. The captain, the naturalists, the patients and the doctors all packed up their belongings, equipment and natural history collections and moved ashore. The naval staff that remained on board, moreover, began regularly to venture beyond the wooden confines of the *Géographe*: officers enjoyed days in town, on occasion the sailors did as well, and teams of men left the ship on a daily basis to collect bread from the bakery, wash the laundry, gather fresh water and complete various other tasks. With such an extension of physical and personal space, with such regular crossings of the line between ship and shore, and particularly with the increased distance from the captain's authority, came a sense of liberty. It was not long before certain sailors and officers were attempting to avoid their responsibilities or, conversely, to assume more power or privilege than that to which they were actually entitled.

During this first week at anchor, however, Baudin took measures to impose an authoritative presence aboard his ship. He put in place a chain of command with himself clearly at its head, closely supervising shipboard affairs, and he demanded that at least two senior officers remain on deck at all times.[6] These measures also served more broadly to preserve what John Gascoigne refers to as the 'lilliputian polity',[7] or hierarchical ordering, which was a vital part of shipboard life and which, during the sojourn, was disturbed by the many absences and the regular comings and goings between ship and shore.

This framework of authority and responsibility was supported by regulations that Baudin designed specifically for the Port Jackson stay. An expedition at anchor for five months in British territory required different rules from those of an expedition at sea, and a copy of Baudin's instructions from the Minister of Marine, which was kept in the logbook of the *Géographe*, authorized Baudin to handle this situation using his discretion. It stated:

> The First Consul, considering the sole way to achieve complete success on your glorious expedition is to oblige you to follow only those laws and regulations of the military marine that are in keeping with the circumstances in which you will find yourself, leaves you at liberty to establish on board the ships entrusted to you whatever form of service, regulation and discipline that you believe appropriate to maintain subordination and punctiliousness in the duties each one of those accompanying you will have to carry out.[8]

The rules set out in the ship's log were critical in counterbalancing the negative effects of staying in a foreign port.[9] This was so, not least of all, because they complemented the rules laid out by Governor King and thus served to uphold Baudin's status as commander of the expedition. More precisely, however, they offset the freedom created by the opportunity for excursions ashore by limiting the men's movements, setting constraints on their conduct and restricting the number of expeditioners who took lodgings in Sydney.[10] The regulations were made clear to all on board and in a way that demonstrates the continued significance of different spaces of the ship to status and order: they were presented by Lieutenant François-Michel Ronsard to the officers in the *chambre du conseil*, and to the crew on the rear forecastle.

Further changes to life aboard the *Géographe* concerned ritual, duties and activity. During the sojourn, many of the men's usual daily activities were unnecessary. Baudin quickly established new routines to replace them. The need to refit the ship, careen and repair it and embark and stow supplies, among many other tasks, meant that the sailors, midshipmen and officers were almost constantly occupied. Instead of a collapse in routine and busyness there was only a brief period of transition to new activities.

Baudin no doubt had in mind the previous ports of call at Tenerife, Mauritius and Timor. At each stop the boundaries defining the shipboard world had faded further and, at the same time, naval discipline had been further submerged under presumptions of social status and privilege. With an imperfect survey of the south coast behind him and, realistically, his last opportunity for a new and better voyage at hand, Baudin could not afford that final step toward mutiny. He made a concerted effort to counteract the ambivalent nature of life on the *Géographe* in port, reminding his men that he, and not Governor King or either of his lieutenants, remained in command while establishing clear boundaries and a distinct shipboard lifestyle.

Below Deck

The crew constituted the largest part of the ship's complement and was responsible for the innumerable everyday tasks without which the ship could not sail. Of all the members of an expedition, the sailors were typically the most experienced in the life of seafaring. At sea, order amongst their ranks was rarely an issue. As Nicholas Rodger points out, a collective understanding of the necessity of cooperating and obeying commands was a matter of survival there and owed little to the authority of the officers.[11] However, the crew in port was notoriously different to the crew at sea. Mary Conley explores how the British 'Jack in Port', for his part, appeared in ballads, theatre and the popular imagination as a 'drunken carouser' and a 'sexual menace';[12] in fact, this caricature had some basis in reality and could apply just as well to many French sailors as to their British counterparts. During the Port Jackson sojourn, Baudin received a letter from two women, seemingly prostitutes, of Sydney. They complained that a pair of drunken French sailors had broken into their house and, causing a commotion, broken some of their belongings. Perhaps the commander ordered the men punished and even compensated the women for the damage; however, the expedition's papers provide no further mention of the incident. This quiet response stands in contrast to that concerning the thefts of cloth and gunpowder, recounted in the previous chapter, and to what would probably have been the response had the perpetrators been officers. As Christopher Forth points out, belief in degrees of masculinity and sensibility endured in French society despite the social reforms of the Revolution and lower-class men, like sailors, were believed to embody a more raw masculinity and a coarser sensibility than men higher up the social scale.[13] Nevertheless, many French sailors had been politicized during the early 1790s by the reforms of the French Revolution and one might therefore expect their behaviour in port to have been affected by their political opinions as much as by their access to rum and women. At Port Jackson, the sailors aboard the *Géographe* were supervised closely and their conduct, particularly their misconduct, was one of the most common topics of discussion among the officers. This was only natural, however, given that it was their shared supervision of the sailors and their affairs which largely filled their days aboard the ship. Midshipmen oversaw most of the sailors' daily tasks while the senior officers, under the close supervision of the commander, were mainly responsible for their behaviour and well-being. It is by examining precisely how this line of authority functioned and how order was maintained among the sailors during these months at anchor, that one gains a rare glimpse of Republican naval and social relations below deck.

The clearest view is obtained via the daily entries made in the logbook of the *Géographe*.[14] And what appears in the foreground of this view is the matter of insubordination, shadowed by the significance to the sailors of official naval hierarchy. The reforms of the Revolution had accelerated a move toward meritocratic practices and imposed Republican ideology but it had not altered the traditional hierarchical structure of the navy. The endurance of this rigid chain of command during comprehensive revolutionary change was no doubt a comfort to sailors. It had always provided the framework for a sense of safety, as mentioned above, as well as identity. At Port Jackson, however, while Baudin did put in place a chain of command aboard the *Géographe*, it was a slightly different one from that which had operated previously during the course of the expedition.

The two sub-lieutenants aboard the *Géographe*, Gaspard Bonnefoy and Joseph Ransonnet, although they did not hold the highest rank, were put in charge of the ship while on guard duty. They had not long been promoted from the rank of midshipman and at Port Jackson they took charge of the ship, in Baudin's absence, for the first time. Their authority in this role was clearly respected less by the sailors than that of the lieutenants, Ronsard and Henri Freycinet. The sailors often refused to carry out the tasks given to them by Bonnefoy and Ransonnet: they argued, occasionally they went to Sydney without permission and sometimes they stayed in town for two or three days until they were found or, exhausted, they returned voluntarily. On one occasion, petty officer Jean-Pierre Billard lost his temper and punched Ransonnet in the face.[15] To make matters worse, there was evident discord between the sub-lieutenants and their immediate superiors. While that relationship will be discussed further on, one example may be observed here. The officers, when on guard duty, typically restrained disobedient sailors by placing them in irons for a set period of time; however, Ronsard occasionally took it upon himself to release to sailors restrained by Ransonnet or Bonnefoy.[16] In this way, he undermined the authority of the sub-lieutenants with respect to their subordinates.

These sorts of tensions, which occurred in relation to situations of authority during the Revolutionary or post-Revolutionary era, are often seen as expressions of class conflict. However, neither the Revolution nor the turn toward authoritarian rule under Bonaparte was truly about class in itself. There is certainly nothing to suggest that the insubordination of Baudin's sailors at Port Jackson was related in any way to class prejudice. Bonnefoy's family was aristocratic, but so was Henri Freycinet's. Neither is there any evidence to suggest that the sailors' respect for their superiors was affected by loyalty to the Republic: although Ransonnet was Belgian, Bonnefoy was French, and the sailors demonstrated no preference for Bonnefoy over Ransonnet. Instead, what the circumstances indicate is that the sailors perceived a weakness in the line of authority – a weakness which disturbed their sense of stability and presented opportunities for escape.

Undoubtedly, the sailors' tendency to challenge Bonnefoy and Ransonnet would have come to the attention of Baudin, given that he was shown a daily report of events on board. However, there is no indication that the commander tried to resolve the situation. It is possible that Baudin chose not to step in on behalf of his sub-lieutenants because it might discredit them further in the eyes of the crewmen. He may have wanted Bonnefoy and Ransonnet to use these opportunities of being in charge of the *Géographe* to develop their sense of authority and to improve their leadership skills. As Dening points out, it was important that a commander allow his men to 'find their own levels of authority independent of his'.[17] Bonnefoy and Ransonnet would eventually be shown greater respect from their subordinates if they earned it independently.

In any case, none of these incidents – even the sailors' drunken jaunts in Sydney after they received their pay – had significantly disrupted the running of the ship. There were few serious incidents of misconduct and, as the sojourn drew to a close, it was only crew members, of all the men on the expedition, who received promotions.[18] Sailors experienced the most restrictions and the least privileges aboard the ship and, consequently, while they did typically make the most of their opportunities for recreation, they also appreciated fair and comfortable conditions. Throughout the sojourn, in addition to setting clear boundaries and ensuring that the sailors were kept busy, Baudin remained attentive to the well-being of his crew. The navy had long acknowledged that sailors were a precious commodity but, since the 1780s when the Marquis de Castries introduced his *Code*,[19] naval authorities had made the improvement of their conditions a priority and, no doubt, Baudin had been influenced by recent reforms. Despite his lack of funds, he purchased new clothes for the expedition's entire crew, including yards of scarlet cloth and serge;[20] believing in the importance of allowing sailors time for relaxation when in port, he sometimes ordered groups of sailors to be sent ashore for a day and he gave them money to spend during their recreation time;[21] and, he made medical care one of his highest priorities. Early during the sojourn, Ransonnet was charged with the responsibility of inspecting the sailors and informing them of when they needed to wash themselves, as well as deciding when it was time for them to clean their laundry.[22] And the crew found no need to complain about the quality or quantity of their daily provisions. Baudin was evidently conscious, as he had been during previous stopovers, of the fact that, as Dening puts it, 'when it came to captains who were also pursers, sailors' stomachs were also spaces of power'.[23] Finally, although the officer in duty was generally responsible for managing discipline, Baudin – determined to ensure that punishments were just and reasonable – prohibited the officers from imposing any punishments without his direct order. When misconduct occurred, the sailor at fault was to be restrained by the officer on duty and the incident reported to Baudin. Occasionally, the officers also received further orders instructing them to impose or end particular punishments.[24] How discipline was enforced was probably the most critical issue when it came to the smooth functioning of a ship and, at Port Jackson, it was clear that the commander had the issue firmly in hand. Some of the sailors clearly had difficulty and perhaps resented being required to work with the temptations of Sydney Town within sight and to conform to an untraditional line of authority. However, as Gascoigne notes, seamen would typically tolerate greater hardships than that as long as they were understood as part of an equitable moral economy.[25]

On the topic of shipboard order and well-being, however, it is worth briefly extending the definition of the 'crew'. It is often forgotten that sailors did not in fact hold the lowest status aboard a ship, and that it is not actually their story that is the least understood – there were slaves aboard the *Géographe* as well: what was their position in the context of this moral economy of the lower deck? Unfortunately, they are barely visible in the records of the expedition: correspondence reveals that two slaves from aboard the *Naturaliste* were given by Hamelin to Governor King as a gift during his first sojourn,[26] while those aboard the *Géographe* rated a mention only when the issue of misconduct and punishment arose. In late August, Henri Freycinet reported in the logbook that '*le nègre* Hervé, having suddenly mistreated and brought bloodshed to *le petit noir* in the service of the officers, had been tied in a cross on the shroud for two hours'.[27] He added that Hervé 'would not submit to any other punishment'; however, Ronsard, in his journal, explained that it was because of this particular disciplinary action that, the following day, Baudin had 'expressly forbid any punishments to be inflicted on board without the direct order of the commander'.[28]

Although kept as slaves aboard the French ship, then, Hervé and the child were treated by Baudin with the same sense of justice, in regard to disciplinary measures, as the other members of the expedition. They clearly held an ambiguous status on the *Géographe*. The expedition had not initially included any slaves – when the ships had farewelled Le Havre in 1800 slavery in the French Empire had been abolished for six years. Baudin had taken these individuals on board at Mauritius, where colonial administrators were stubbornly ignoring abolition. By the time that he reached Port Jackson in the winter of 1802 slavery had been re-established in the French colonies by Bonaparte and Toussaint L'Ouverture, leader of the Haitian Revolution, had been imprisoned in France. Baudin's expedition had sailed across the various phases of French abolition and re-establishment of the slave trade. The incident at Port Jackson, then, presents a profound image: much to the commander's displeasure, an African slave tied to the shroud of a French scientific vessel in a British colony – there was encapsulated the Republic's struggle with the concept of human rights.

Below deck, more generally, concerns about rights and privileges were expressed in a way that reflected a combination of traditional priorities and desires that were very much of their time. As French sailors and citizens, these men were probably no less political or patriotic, at base, than they or other crews had been during the Revolution, but in 1802 they answered to a strong central government rather than the ambiguous authority of popular sovereignty.[29] Unlike the mutinous crews of the 1790s, Baudin's sailors did not rebel against privileged command but demanded a firm and traditional demonstration of authority. They were less concerned about political ideals than about security and stability – as were most of their compatriots at home at this time.

The Quarter Deck

It was handling the *Géographe*'s small group of officers rather than supervising its large crew that most occupied the commander during these months in port. The senior naval staff played a central role on the ship and in the expedition more generally. They were in an almost parental position of authority over the sailors and, at the same time, were managed in a paternal manner by Baudin. Moreover, their status and responsibilities overlapped with those of the naturalists. They shared with the naturalists the privileged spaces of the quarterdeck, while at sea, and the captain's table both at sea and in port. Their affairs were at the heart of shipboard politics and could accordingly have significant consequences for both the crew and Baudin if not handled effectively.

It has been argued that Baudin did not in fact manage the officers well during these months; indeed, Frank Horner asserts that the commander's efforts to command the *Géographe* from ashore were 'hopeless'.[30] He claims that there was no recognized line of authority and sets a scene of overall disarray and discontent among the officers during the Port Jackson stay. Horner's analysis, however, is based almost entirely on the journal entries of Lieutenant Ronsard and is not contextualized in the history of the French navy; it therefore gives a rather one-sided and superficial view.[31] There were tensions on the quarter deck, but they relate more to contemporary naval culture and politics than the failings of an individual commander.

Baudin's senior naval staff belonged to a new Republican officer corps. The *Grand Corps* of the *ancien régime* – traditionally composed of noble officers and based on authority derived from aristocratic privilege – had disintegrated in 1791, the 'year of military emigration'. Those few who remained, one might assume, were particularly devoted to their careers and possibly, too, to service for their nation. Upon the declaration of the French Republic in September 1792, Gaspard Monge, Minister of Marine and the Colonies, had certainly been optimistic. In a report addressed to the National Convention, he declared: 'Today, the Minister assures the Convention that it will find in the national navy a nursery of fine navigators, capable of defending the Republic's flag and devoted to the maintenance of its liberty'.[32] 'Nursery' was certainly an apt title for the officer corps as it then stood. Only days before this report was presented, the Legislative Assembly had introduced emergency legislation designed to expand the corps as quickly as possible. It was in fact only four months later that, at the ages of 13 and 14, Louis and Henri Freycinet were driven to Toulon by their father and enlisted as midshipmen in the navy.[33] The following year, the Convention approved a campaign to purge the naval officer corps and ensure that it would be composed of only 'true friends of the people'.[34] André Jeanbon Saint-André, representative of the Committee of Public Safety, led an army of government agents in gathering on every officer 'observations, information, facts' that would lay bare their political loyalties: 'no enemy of the People, no equivocal or doubtful man, was admitted when we could tear away the mask behind which he was hiding', asserted Jeanbon.[35] Of Baudin's officers, only Ronsard was associated with the navy at this time and the report attesting to his suitability for appointment is still available in his personal dossier: 'his Patriotism: *pur et juste*', it states, 'Observations of his moral conduct: irreproachable', 'idem of his political conduct: laudable'.[36] Thus, Ronsard successfully navigated the construction of a new officer corps and went on to serve under authorities that sought not only to revolutionize the navy but to devise a naval effort to defend the Republic. In 1793, the Committee proposed a decree to hasten the reconstruction of the Mediterranean fleet and, as he introduced it, Bertrand Barère declared: 'Let us build ships, and let us forge arms. To the dockyards, citizens! To the workshops! This is the cry of the Republic!'[37] Of course, the navy claimed no great triumphs against France's enemies either during these years or later during the administration of the Consulate. Perhaps one of the ideas that motivated Bonaparte to support the Baudin expedition was that this venture could exhibit the new navy and boost naval morale in France; certainly, the expectations and ambitions of Baudin's officers, as well as the tensions between them, must have been shaped by this revolutionization of the fleet.

Although these officers all represented the new, Republican, navy, and must all have been conscious of that circumstance, by no means had they followed the same path to the *Géographe*. For example, Ronsard – 19 years old when he joined the navy and 20 when, one year later, the Revolution began – had been through each phase of the Revolution; however, while the Committee's report tells one story of his position concerning the event, a letter written during the later Restoration years, which states that Ronsard had never been 'a man of the Revolution' and describes the persecution of his Royalist family members as well as his own imprisonment during the Terror, tells quite another.[38] For his experiences and, quite possibly too, his political principles, Ronsard stood in particular contrast to his companions on the quarter deck at Port Jackson. Altogether, the officers were of diverse backgrounds with correspondingly, if only slightly, varied political leanings, ambitions and standards of masculinity and sensibility. Still, all but Ronsard had joined the fleet as adolescents between 1793 and 1796. They were the fresh, young recruits, the replacements for the officers of the *Grand Corps*.

The freshest among them were the midshipmen: the apprentice or junior officers. It was among these young men that the least disciplined staff aboard the *Géographe* could be found; however, they seem neither to have caused much disruption nor to have played a significant role during the expedition's sojourn at Port Jackson. At this stage of the voyage, there were just three midshipmen aboard the *Géographe*: Joseph Brue, who was transferred to the *Naturaliste* immediately following its arrival at Port Jackson in late June; Hyacinthe de Bougainville, transferred towards the end of the sojourn; and Charles Baudin, the only midshipman retained on the *Géographe* for the second campaign. The fact that these men were rather quiet at Port Jackson is of particular note given that, throughout the voyage from France, they had been the cause of considerable frustration – as midshipmen typically were on such expeditions. Moreover, during their time in the British port, Brue and particularly Bougainville showed a clear lack of commitment to the objectives of the expedition. Bougainville requested early during the sojourn to be transferred to the *Naturaliste* so that he could return to France. When Baudin refused, he devoted the remaining period to feigning illness and thus spent much of the sojourn in hospital and out of mischief. He was the son of Louis-Antoine de Bougainville – aristocrat and celebrated naval voyager. He had experienced a privileged upbringing and it is quite possible that, despite the egalitarian reforms of the Revolution, he took the possibility of a successful naval career for granted. Indeed, his position under Baudin was his first in the navy and it had been obtained at the request of his father. By contrast, although the commander did need to discipline him on occasion, Charles Baudin appears to have been dedicated to the expedition. He behaved and worked well; he was even given an opportunity to act as the officer on duty while at Port Jackson. This midshipman, it seems, did not take his position on the expedition for granted. His father, Baudin des Ardennes, president of the *Conseil des Anciens* in 1795, had died before the end of the Revolution and Charles and his mother had lived for some time in poverty. When Charles joined the navy in 1799 it was at the suggestion of the First Consul, an admirer of his father. At Port Jackson, by all accounts, Charles was pleased to be continuing the expedition while Bougainville was eager to return home; this was for all of the midshipmen aboard the French ships, according to Charles, 'truly a happy time'.[39]

The same might not be said for the senior officers; that is, if the journal of Ronsard alone were to be relied upon. On the basis of Ronsard's journal entries, Horner blames Baudin's refusal to delegate his authority to either of his lieutenants for a state of disorder aboard the *Géographe* at Port Jackson. However, it is not certain that this arrangement caused confusion among the senior officers. Baudin made their rights and roles as individuals and in relation to each other clear through what Horner condemns as a 'steady stream of orders and complaints':[40] He established a list designating to each senior officer responsibility for a particular part of the ship, an explanation of the duties of each officer on guard duty, and some detailed supplementary instructions that were based on his observations of the officers' conduct over time. He also recognized the difference in rank between the lieutenants and the sub-lieutenants by authorizing Ronsard and Henri Freycinet to give orders to Bonnefoy and Ransonnet as well as instructing them to report back to him on how well those orders had been obeyed. Each of these orders was communicated in writing and posted by the officer on guard duty in the logbook of the *Géographe*. Each senior officer had his own, distinct, space of authority aboard the ship and it was clear that Baudin himself ultimately remained in command. This arrangement did not cause problems amongst the midshipmen or the crew and there is no reason why, in itself, it would spark arguments amongst the men or resentment towards Baudin. What may have played a more significant part in the contentious situations that nevertheless arose from time to time was the attitude of each officer to his own position, to his peers and to the expedition.

One altercation of interest in this respect occurred on a September evening during dinner in the great cabin, between Bonnefoy and Henri Freycinet. As recorded by Ronsard, Bonnefoy had derided Henri's friend, Péron, as 'that peasant, your corporal Péron', to which an enraged Henri had retorted: 'a corporal like him is worth 10,000 officers like you!' As tempers rose, Bonnefoy argued further that Péron had insulted him: 'he is a pig', he exclaimed – 'it is you who is a pig!' returned Henri. When Bonnefoy continued to criticize Péron, Henri Freycinet stood upon his rank and ordered him to his cabin. However, arguing that the lieutenant had no right to punish him for an affair that did not concern the service, Bonnefoy refused to obey: 'I hope, sir, that you intend to inform the commander of the motives for which you have punished me', he remarked.[41] In the end, when fully informed of the details of this episode, Baudin concluded that Henri Freycinet and Bonnefoy had been almost equally at fault: Henri, for engaging in a dispute on board, and Bonnefoy, for not knowing when to be silent and for disobeying an order. He also blamed Ronsard, who had been the officer on guard duty at the time, for not intervening to resolve the situation:

> As officer on guard duty ... it was your duty, and you knew it, to impose silence on each man or at least to invite them to go and discuss elsewhere a matter that was absolutely foreign to the service of the ship and of pure opinion.[42]

This incident perhaps seems trivial, at first glance, but it is not. On one level, it highlights the absolute importance in the shipboard world of a professional protocol that separates personal feelings and opinions from issues pertaining officially to the running of the ship. While on deck, the officers were allowed not a moment's reprieve from the roles prescribed to them by their naval rank and responsibilities. It was only on shore that, to some degree, they were permitted to give vent to their emotions. However, even in the rigid environment of the quarter deck, sentiments could not of course be repressed entirely and those that most often emerged were characteristic of naval culture. On another level, then, this episode also reveals the nature of the *Géographe* as what Barbara Rosenwein would call an 'emotional community'.[43] Bonnefoy had been insulted by Péron (one wonders what he had said), but the animosity he felt towards him evidently went deeper – it was rooted in his perception of Péron as an outsider. When Bonnefoy referred to the voyager-naturalist as a 'corporal' he identified him by his association with the army, an institution outside the introverted world of the navy and even unrelated to the expedition. He also ridiculed him by simultaneously pointing to his inferior rank and social class – 'that peasant, corporal

Péron' echoes and inverts the title '*le petit caporal*' (the little corporal), which soldiers of this era often applied affectionately to Bonaparte. Such snobbery reeked of aristocratic prejudice and, of all on the quarter-deck, would have rankled Henri Freycinet most particularly; though of noble background himself, he and his brother Louis are thought to have been freemasons and,[44] if indeed they were, then they would have been deeply committed to principles of liberty and equality. Moreover, both Péron and Henri Freycinet were enthusiastic supporters of the Revolution and the First Consul; therefore Bonnefoy,[45] whose feelings were apparently more ambivalent, was able to attack the lieutenant's sense of loyalty not only to a friend but also to his hero and nation.

Of course, this dispute was exacerbated as well by the questioning of authority and regulation and the boundaries that they formed. It was critical that naval staff refrained from challenging these boundaries, exercised the authority assigned to themselves and respected the authority with which others had been charged. As Baudin pointed out, not only had Bonnefoy and Henri Freycinet in different ways pushed the limits of their positions but Ronsard had failed to fulfil his duty as the officer on guard duty.

Why he remained passive on this occasion is uncertain, but it may be that Ronsard felt out of his depth in this argument. Although it developed into a challenge to authority, the dispute between Bonnefoy and Henri Freycinet, both aristocratic young men, was essentially about personal pride and involved a genteel degree of masculine sensibility with which Ronsard was probably not comfortable engaging. Certainly, the engineer-lieutenant was usually not backward in taking a stand during these months. His anger, though, was typically sparked less by personal insult than by professional questions of rank, authority and career progression.

As noted earlier, Ronsard had shown a tendency during these months to disregard and undermine the authority of the sub-lieutenants working alongside, if ranked below, him. His habit of referring to Ransonnet and Bonnefoy in his journal by their rank alone, 'sub-lieutenant', rather than by their name, or even, in the case of Bonnefoy, because he had been temporarily demoted, 'midshipman' rather than 'officer' on guard duty, further expresses his disrespect for them and his resentment at having to share responsibility for shipboard affairs with men of inferior rank. In mid-October, the situation came to a head. Bonnefoy was the officer on guard duty and Ronsard, responsible for stowage, was supervising a team of sailors, including the baker, as they stowed campaign biscuit in the hold of the *Géographe*. Baudin had made Ransonnet responsible during the sojourn for ensuring that the baker went ashore each day to check on the production of biscuit at Sydney's bakery; however, on this day, Ronsard order the baker to remain on board to help him. He sent the baker to inform Ransonnet of the change and, recognizing that the lieutenant had the right to issue orders to him,

Ransonnet assented. The problem was that he did not attempt to inform Bonnefoy, who was expected to ensure that all men carried out their regular duties – including the baker – and so later ordered the baker to head ashore as usual. When the baker returned to the *Géographe*, Ronsard, furious, confronted him on the bridge and loudly reproached him. Within hearing of others aboard the ship, he declared that the baker had been wrong to obey the officer on guard duty. There followed a vehement war of words in the logbook of the *Géographe*: Ronsard did not believe that he had been obliged to keep the officer on guard duty informed of the change to the service, and spitefully reminded Bonnefoy of his demotion, while Bonnefoy pointed out that in the absence of an order to do otherwise he had only followed regulations. As the officer on guard duty, the sub-lieutenant did need to be aware of what was happening aboard the ship and Ronsard, by ignoring him and publicly admonishing the baker for obeying him, clearly undermined his authority.

Baudin reproached Ronsard angrily and, indicating that for some time he had been observing not only him but Henri Freycinet as well, criticized the leadership of both lieutenants. Although, interestingly, Ronsard and Freycinet managed to share without conflict the most senior position on board, neither of them, especially Ronsard, dealt well with the authority with which Ransonnet and Bonnefoy had been charged and they clashed most often with Bonnefoy. The records of the sojourn give no indication as to why this was so but the very different backgrounds of the sub-lieutenants allow some speculation. Ransonnet originated from Liège, Belgium, and had been employed in volunteer service there before moving to France in 1793 – the year of the National Convention's *levée en masse* and following the French occupation of Liège – to join the French army. He served as his father's *aide-de-camp* until M. Ransonnet's death in 1796, whereupon, after arranging his father's affairs and seeing to the education of his younger brothers, he joined the navy and served for two years before heading to Australia with Baudin.[46] Ransonnet had risen through the ranks and may thus have been seen to represent the political values and militarized masculinity of the Republic. Bonnefoy de Montbazin, however, had an aristocratic background.[47] To what extent his family's standing or wealth was affected by the Revolution is not known but, judging from his scornful remarks about Péron, he still harboured a sense of social superiority. Furthermore, this may have been augmented by a feeling of professional advantage: he had served under Baudin on a natural history voyage before – aboard the *Belle Angelique*.[48] If Bonnefoy did generally demonstrate a degree of social and professional superiority it would have irritated Ronsard in particular. Ronsard had determinedly pursued his career through the midst of the Revolution, and now a young aristocrat 10 years his junior was biting at his heels.

In fact, by all reports, there was no discord between Bonnefoy and Ransonnet; instead, the tension was between the ranks of the senior officers and, moreover, it was weighted most heavily toward Ronsard. The naval engineer was not committed wholeheartedly to the expedition. He was impatient for a promotion and thus cared more about questions of status and codes of honour than the pursuit of knowledge – the five months at anchor passed slowly for him. When Baudin refused to act upon the 'wishes or whims', as he called them, of Ronsard by promoting him, Ronsard threatened to quit the expedition and focus upon his naval career, pronouncing:

> I can go no further without having navigated on ships of the line, thus, the two years that I would pass in this second part of the expedition that you are going to undertake will be given entirely to my education and not at all to my advancement.[49]

This single-minded attitude could well have jarred with the more intellectually oriented officers. It certainly influenced Baudin's treatment of Ronsard; as the commander explained, Ronsard's aspirations 'could not be reconciled with my duties or with my responsibilities toward those who could be compromised by such ill-considered reasoning'. Therefore, while he temporarily appeased Ronsard's agitation by issuing to him orders that might ordinarily have been issued to a second in command, Baudin stopped short of granting him the broader, official, responsibility due to a first lieutenant. Had he delegated authority to an officer whose attitude was not in line with the aims of the expedition, the reasonable level of order aboard the *Géographe* could have been jeopardized.

Not only did Ronsard feel that the second campaign would be a waste of his time but he believed too that Baudin was failing to treat his lieutenants with the respect and honour they deserved as naval officers. With great disloyalty to his captain, he declared in his journal: 'according to regulation 65, a captain who failed to treat his officers with the respect and decency that must always reign between servicemen, was declared incapable of commanding the King's vessel'. This was a serious accusation: honour had long been a critical part of naval, indeed military, custom, and Napoleonic culture had heightened its significance even further. However, Ronsard confused honour with privilege. He deemed himself and Henri Freycinet to be above Baudin's censures, restrictions and directions. Both lieutenants wanted greater freedom to enjoy life on shore and authority to run the ship as they wished. Note, too, that Ronsard was harking back to the days of the 'King's' fleet – a further hint that he may not have been the 'true friend of the people' that the Revolutionaries thought he had been and thus additional cause for his ill fit in the officer corps of the *Géographe*.

In any case, despite Ronsard's bitter complaints, the officers seem to have enjoyed a comfortable sojourn. They were certainly subject to stronger supervision and tighter regulations than they had been during previous stopovers – after the almost mutinous misconduct of his then second in charge at Timor, Baudin had decided to change his style of command and, at Port Jackson, this approach was reinforced by regulations imposed by Governor King – yet the commander provided well for them. Despite the shortage of funds, he paid them for the first time in the voyage. They were able to purchase personal luxuries and to dine well: unlike previously, there were no complaints about the officers' table. As we shall see in the following chapter, Baudin also made a point of firmly defending the honour of his officers when accusations were made against them by their English counterparts. On the part of the commander, in all, a care to the officers' sense of honour and gentlemanly sensibilities as well as enforcement of clear boundaries and a sense of accountability were critical in the maintenance of order on the quarter deck.

Shipboard order had been threatened, to a degree, by Ronsard's preoccupation with the pursuit of a traditional military career, above all, but also by political and cultural differences between his fellow senior officers; thus, a tight command was certainly crucial. Yet order was also largely facilitated by the parallel circumstances of the expedition and the new Republic. As France was entering a new era under the leadership of the First Consul, at Port Jackson, one voyage had drawn to a close and another was beginning. Those junior officers whose discipline and motivation had waned could return to France without dishonour and take up other opportunities in the fleet. In time, Hyacinthe de Bougainville, for example, would command his own maritime expedition and gain the rank of rear admiral. These young men were not troubled by the sense of urgency that drove the older Ronsard but instead were satisfied with their achievements thus far – participation in a South Seas expedition was an important addition to any naval record – and confident about their future prospects. At the same time, those officers who were committed to the mission at hand were facing an opportunity at least to further their experience in maritime exploration if not to avoid returning home having barely fulfilled the hopes of the government and the Institut, and instead to be part of what might become a national triumph.

Commanding from Ashore

The pivotal nature of this sojourn for the course of the Australian voyage, the opportunities it provided to the expedition members and what was required for these opportunities to become possible, were all made clear by Baudin's manner of command. True, Baudin, himself, was not a physical presence aboard the *Géographe* for the most part of these months; yet his authority there was nonetheless tangible. His close supervision of the crew and the officers, via reports, letters and written orders as well as meetings on shore, has been demonstrated. His authority remained evident also through the maintenance of the commander's personal space aboard the ship and, correspondingly, the limitation of the officers' presumptions. In an outline of the responsibilities to be assumed by the senior officers, Baudin declared, for example, that 'the two wardrobes in the great cabin were not intended for the usage of Messieurs Taillefer and Bougainville and even less as their *cabinet de toilette*. All objects that belong to one or the other are to be removed'.[50] His command was made palpable, too, by the resolute aspiration that shaped and directed it across the shore.

This Australian expedition was an exceptional opportunity for Baudin. It was an opportunity that had been opened up to him not only by his own efforts and achievements but by the reforms of the Revolution. For years Baudin, being of non-aristocratic birth, had served in the French navy as an auxiliary officer, an *officier bleu*. This was a temporary status dependent largely on connections and good fortune. To gain employment during the late 1780s, he had resorted to sailing Austrian vessels and then had to prove his patriotism to the Republican government before, finally, being appointed to the French officer corps in 1798. In his efforts to move upward in his career at sea, despite his social class and limited formal education, he had led a series of botanical voyages; but this voyage was by the far the most important and, for Baudin, its success was critical. This was physically the largest expedition Baudin had led, comprising two vessels and over 200 men, and it was the largest in terms of expectations as well. The Australian voyage was intended to continue the tradition of grand voyages of discovery set by Bougainville, Cook and La Pérouse, but, simultaneously, to effect a major turn in that tradition, by narrowing and deepening the investigative focus of maritime exploration. In 1802, these expectations remained unfulfilled and one way in which Baudin sought to remedy that was to renew his command.

To this end, he returned to the instructions issued to him by the Minister of Marine, Pierre-Alexandre-Laurent Forfait. Forfait's first point was that Baudin should be 'both leader and father' to his men. This reflected a traditional belief that naval captains should demonstrate a paternal vigilance by guiding and taking responsibility for the education, welfare and discipline of their men. However, the warmth normally associated with the relationship between a father and his children was expected to be limited in this context and tempered further by being combined with the role of leader. On deck, the degree of familiarity between a captain and his men varied, depending particularly upon the size of the expedition and the make-up of the complement, but a degree of distance was generally expected. That said, as Revolutionary ideals of equality were embraced and, later, as the French army became 'Napoleonic' – in that it was influenced by notions of manly friendship between ranks and by a perception of Bonaparte's paternal care and leadership[51] – such an expectation could not be assumed. Social equality had been introduced but the smooth functioning of naval companies still depended upon strict conformity to a hierarchical system. The tension this situation created was still being worked out and it made for potentially fraught emotional environments aboard the vessels of the fleet. In his memoirs, Charles Baudin wrote that he and his companions had been disappointed upon first meeting Baudin. 'Like spoilt children', admitted Charles, they had been ready to embrace their commander, to 'eat from his hands', but Baudin had rejected their advances and sought instead to establish discipline and hierarchy. Indeed, the two would go hand in hand in Baudin's command of this expedition. During the course of the outward voyage and the first campaign, when discipline had been allowed to slacken, the hierarchical ordering aboard the *Géographe* had slipped as well – disorder and conflict resulted. The maintenance of discipline and hierarchy, and the certain distance from his men that this maintenance required, were clearly of the utmost importance to Baudin at Port Jackson. Together, they formed a framework for his role as 'leader and father' and, indeed, for the emotional community of the ship. What this means, more precisely, is that Baudin was responsible for managing not merely the division of tasks and the naval education of his men but, just as importantly, the affective state of his shipboard society. In the advice he gave to Louis Freycinet, upon officially appointing him as captain of the *Casuarina*, it is clear that he believed one's authority as captain depended upon the utmost discipline in this regard:

> Never forget [he wrote] that, if it is difficult to command men, it is nevertheless glorious to drive them well. One succeeds almost always when one conducts oneself toward them with moderation, prudence and justice. According to these principles, which are not – or very rarely – put into practice by young officers, you must avoid all occasions that could compromise your authority and even your person. He who commands is not to be excused when, indulging in excesses that the law forbids, he listens only to his passions and arbitrarily or capriciously punishes an individual who has merited disciplinary action.[52]

This approach corresponded closely with the wishes of Forfait. The Minister's instructions had directed Baudin further to 'let no prejudice influence the acts of authority' he performed, and, in conclusion, he had added: 'I repeat to you once again to maintain good order and strictness in compliance with the form you have adopted for the service. The slightest negligence on this point could have the most fatal consequences'. Baudin's conformity with these directions at Port Jackson was evident particularly in his resistance to Ronsard's demands for promotion as well as his determination not to delegate his authority to either of the lieutenants. These decisions put Ronsard offside with his commander, but they were key to upholding the hierarchy of authority and the order that that brought to the expedition. Dening explains, in his anthropology of an expedition, that such delegation as that demanded by Ronsard tended only to decentre the system of discipline aboard ships.[53] In fact, this is what occurred on the *Recherche* in 1792 when d'Entrecasteaux, yielding to the constant pleas and presumptuous conduct of his ambitious first lieutenant, delegated too much authority; before too long at sea, Bruni d'Entrecasteaux was no longer regarded as the leader of the expedition.[54] Little imagination is required to see that, had a lieutenant been given responsibility for the *Géographe* during the five-month stay at Port Jackson, while the commander remained ashore, the balance of power would most probably have shifted and altered, to its detriment, the system of discipline among the expeditioners.

It is interesting that Forfait's instructions and, especially, Baudin's style of command, bear little resemblance to the leadership of d'Entrecasteaux, who sailed the Australian waters just ten years earlier, during the Revolution. This renowned voyager, whose expedition, as indicated, was plagued by blurred hierarchical boundaries and shipboard conflict, is said to have been 'too easy-going, too good even' and 'afraid of attracting the enmity of the officer corps, to whom he [was] too attached'.[55] In fact, one needs to look back to the height of the Enlightenment era, and the command of James Cook, to find a distinct similarity. What made Cook such a successful captain, explains Gascoigne, was in part his ability to be seen to be in command, which meant being 'a rather distant figure',[56] and which resulted in Cook being loved as well as feared by his men. By comparison, Baudin's standing with his men was more ambiguous – there were more politics on this expedition – but the French commander was clearly attempting to emulate a style of command more reminiscent of Cook than his predecessor d'Entrecasteaux. On his previous, smaller, expedition aboard the *Belle Angelique*, Baudin had been a more relaxed captain and had fostered close friendships with some of his men – though mainly naturalists rather than naval staff. His preference on this Australian voyage for an older, more authoritative, disciplined and hierarchical, leadership largely reflects the greater size and significance of the expedition. But it also corresponds with two significant shifts occurring in France at this time: firstly, the hardening of affective standards – the 'erasure of sentimentality' and the 'disciplining of passions' as William Reddy puts it – which occurred in reaction to a perceived excess of emotion seen to have culminated in the Reign of Terror;[57] and, secondly, the turn away from the social disorder of the Revolutionary years toward a central system of authority and a national sense of stability. These shifts possibly encouraged Forfait in his advocacy of a traditional system of discipline for the Baudin expedition.

At Port Jackson, however, Baudin also had more immediate reasons to tighten his command. The restrictions that he placed on his senior officers and the number of duties he gave to them were greater than usual and he went to some effort to explain his approach. One month into the sojourn, he had criticized the lieutenants and sub-lieutenants for preferring 'passing pleasures to real and obligatory duties' during their time in port. 'The campaign that I am to undertake is not that of a warship, where ports of call are opportunities for pleasures or amusements. On the contrary, it requires more active and laborious work than at sea', he explained. This explanation may have been largely for the benefit of those participating in natural history voyage for first time, but those men had already been engaged on this voyage for over eighteen months by this point and experienced three ports of call. Clearly, Baudin was taking this sojourn more seriously. There was no space, in his mind, for generosity or indulgence this time. The expedition was to remain at anchor for an unusually long period of time so a greater effort was required to ensure that the expedition did not fall into disarray. Moreover, Baudin intended to undertake a fresh, more productive voyage beginning from that port. To do that, he needed to maintain the expedition's commitment and motivation as well as to keep it disciplined and working efficiently. He also needed to safeguard the colonists' cooperation: 'the occasions to do harm are too numerous and too easy in the place and state in which the corvette finds itself to conduct ourselves otherwise', he explained further to his men.[58] The expedition was in a vulnerable position – reliant on cooperation either to go back or to move forward; and the place where this cooperation was required was one which encapsulated Anglo-French rivalries. The expeditioners needed to remain conscious that they were not alone but performing in a colonial theatre. In fact, that circumstance gave Baudin additional motivation to avoid delegating to a second in command. By personally ensuring that his ship's company remained disciplined, orderly and firmly under his control, if disgruntled, he could present a solid and focused unit to the English. And, furthermore, by demonstrating, before his men and before the colonists, the strength of his command alongside King's own authority as governor of the colony, Baudin asserted his authority over the expedition and his role as a representative of the French Republic. His men had been impressed by both King and the colony. Baudin needed to ensure that they did not sway toward their new surroundings and acquaintances and away from his command and the mission at hand.

In part then, it was by returning to his instructions and prioritizing the basic naval principles of discipline and hierarchy – for direct effect aboard the *Géographe* and for performative effect in the space of the British colony – that Baudin attempted, anew, to satisfy the purpose of his voyage and to fulfil the ambitions of his government and the Institut. He had been attempting throughout the course of the voyage to enforce this 'Cook' style of order upon the *Géographe*. The ship heaved, though, with notions of individual rights and possibilities as well as idealistic concepts of leadership. There were moments of tension, conflict even, misconduct and frustration aboard the *Géographe* during these months at anchor. There were also binding commonalities – patriotism and ambition – and these came to the fore as the men found themselves both upon the imperial stage and facing unexpected opportunities.

4 THE FRENCH AND THE BRITISH: A DIPLOMATIC RELATIONSHIP

Questions of honour, authority and status were just as vital to relations between the French and the British as they were to life aboard the ships of the Republic. While, on deck, officers vied for promotions, defended their pride, shared shipboard responsibilities and, at the end of the Revolution, struggled to reconcile equality with hierarchical order, across the Channel representatives of France and Britain fought for ascendancy, asserted their virtues, shared knowledge and, from 1793 until 1802, battled over liberty and empire. British forces had been the natural focus for the new Republican navy and, as the Revolutionary Wars came under the sway of Bonaparte in the late 1790s, Britain itself was threatened. Although the invasion did not eventuate, Bonaparte's Irish expedition of 1796 and Egyptian Campaign of 1798 were direct attacks on British imperial interests and, despite the clear superiority of their navy, the British soon found themselves at a disadvantage, without a continental ally. They were forced to accept the Treaty of Amiens in March 1802, and, still, little had actually been settled between themselves and the French. Just as Baudin did not conclude his mission when he anchored at Port Jackson but prepared for a new voyage, so French and British forces in Europe poised this year for the renewal of war.[1] The 'sciences were never [literally] at war', but relations between the colonists at Port Jackson and their French guests were clearly underlaid by political anxieties – particularly on the part of the British. In the interests of peace between their nations and of the well-being of both the expedition and the colony, these anxieties needed to be negotiated via the utmost attention to diplomatic etiquette.

It was not just war itself that was critical to Anglo-French relations at this point in time, and especially to the relations at Port Jackson, but the particular nature of this war. Under the Directory, the French objective had begun to advance from ideological to imperial campaigns. During the late eighteenth century France had lost considerable colonial territory to the British: possessions in North America as well as India at the conclusion of the Seven Years War in 1763, Caribbean colonies during the Revolutionary Wars. The balance of commercial and geo-political strength had tipped decidedly in favour of the British and, though France had already been searching for new strategic and trade opportunities via a series of Pacific voyages in the late eighteenth century, the First Consul now sought a more decisive remedy.

It was not his intention to use the Baudin expedition to establish an antipodean settlement, but it would have been only natural for the British authorities to suspect otherwise. The escalation of imperial rivalry in Europe coincided with a rising fear, at Port Jackson, that the French might invade or establish a settlement near the British colony. The two nations had been rivalling each other via exploration in the South Seas since the mid-eighteenth century. Each attempted to improve on the claims of those before them – with more extensive charts, larger natural history collections and closer relationships with the Indigenous Australians and Pacific Islanders. In 1788, when French authorities learned that the British were about to establish a settlement at Botany Bay, they directed circumnavigator Jean-François de Galaup de La Pérouse to head there immediately and to investigate its progress. The d'Entrecasteaux expedition followed only three years later, not to visit the settlement but to investigate New Holland more closely, to explore Van Diemen's Land and neighbouring islands, and to search for any sign of the preceding French discoverers. Of course, British authorities were acutely aware of the French shadow over their colonial project; among a host of other considerations, fear that the French would themselves establish a settlement in the region had impelled them initially to colonize Botany Bay and soon afterward to build additional colonies at Port Phillip, Port Dalrymple and eventually the Swan River. Governor King was informed about the Baudin expedition within months of its departure from Le Havre. He must have been mindful of the circumstances: that the last French discovery ships had only recently left Australian waters, that this voyage coincided very closely with Bonaparte's seizure of power and that this was not a round-the-world or even a Pacific voyage but a mission concentrated specifically upon Australia. He knew that the expedition was not intended to visit Port Jackson but expected that it would and,[2] no doubt, he saw reason to be concerned.

Baudin carried a passport from the British government; still, assistance from Governor King was not a matter of course. While France and Britain generally held strong to the tradition of separating scientific internationalism from the business of war, they did so to varying degrees and certain authorities sometimes overlooked it altogether. In 1795, for example, a British naval expedition seized D'Entrecasteaux's natural history collection and, from 1803, the French governor of Isle de France would keep Matthew Flinders imprisoned and unable to publish the results of his Australian voyage for seven years. On the basis that King, for his own part, did abide by tradition and the conditions of the passport, his hospitality is typically understood to have derived from the purest of motives – generosity, humanity, a commitment to the international pursuit of knowledge – and his political concerns are given limited significance.[3] However, there are two problems with ascribing the success of this episode merely to 'generosity'. Firstly, it is inaccurate, for it implies that King, 'nursing' and 'feeding' them, catered to the Frenchmen's every need while, as shown earlier, what he actually did was to allow the Frenchmen to stay in the colony on the provision that they provided for themselves. Secondly, it obscures the intricacies of what in fact must have been at times a rather difficult relationship.

When Baudin arrived at Port Jackson, he did not sail into the open arms of a selfless governor – he entered into fraught foreign relations. Fulfilling his objectives in the British colony was not to be a simple task. It must be remembered that Baudin sought more than just respite and replenishment: he required accommodation until the summer and the opportunity to prepare for a second campaign. He needed to establish trusting relationships with his hosts. At the same time, though, he was also to make a certain impression. The Minister of Marine and Colonies had instructed him, in his capacity as an envoy of the Republic, to 'make France's name honoured in all the countries he visits' – to give foreigners 'an exact idea of present state of France and of the prosperity assured to it'.[4] The means by which he was to accomplish this, the Minister asserted, would be provided by 'the glorious success of our armies, the power and wisdom of the government, the great and liberal designs of the First Consul for the pacification of Europe, and the tranquillity that he has restored to France'.[5]

The contact between the French voyagers and the British colonists in 1802 does not represent a story simply about British benevolence or even about the strength of the 'commonwealth of learning'. There is a more nuanced story of Anglo-French relations to be drawn from how individuals in this encounter interacted before the background of discovery, empire and Australia, in how they drew on common cultural traditions, rituals and rhetoric to manage their fears and to balance their interests. It is drawn too from a closer look at the British view and treatment of the expedition as well as by attending, for the first time, to how Baudin navigated the politics of this episode in the interests of his government. Between the expeditioners and the colonists there was both pragmatism and theatre: they 'acted out their scientific and humanistic selves', they 'jostled to see what the Pacific said to them of their relations of dominance', they 'vied in testing the extensions of their sovereignty and the effectiveness of their presence'.[6] As shown, however, this was also an encounter in place and time, and, moreover, an encounter between individuals. 'The civilized' who met at Port Jackson in 1802 were not agents of generic European, French or British imperialism.[7] They were colonial authorities and voyager-naturalists, naval officers, men and patriots of their era and they met in a contact zone where, combined with politics, certain values and standards integral to contemporary French and British cultures significantly shaped the host–guest relationship.

The Baudin Expedition through British Eyes

To begin with, there were intentions and expectations on either side. Baudin approached the colony prepared, as noted above, to represent France in a certain manner; however, the British had their own preconceptions of the French nation and of its interests. The expedition as a whole and this colonial episode in particular took place within an intricate web of cosmopolitan ideals and imperial ambitions, at the centre of which was Joseph Banks, President of the Royal Society in London, corresponding member of the Institut National and leading figure in the colonization of New South Wales. His association with the French expedition was significant in a number of ways – not least through Governor King and Matthew Flinders; ultimately, it both facilitated and complicated Baudin's mission at Port Jackson.

For Banks, more explicitly perhaps than for other naturalists of his time, science was primarily for 'the service of empire'.[8] As a member of the Republic of Learning, it was Banks's duty to provide essential support to French voyages of exploration and, accordingly, it was he who helped the Institut National to obtain British passports for the *Géographe* and the *Naturaliste*. His distrust of the French and his concern for British interests, however, soon came to the fore. Just two months after the expedition's departure, Banks wrote to the Earl of Spencer, First Lord of the Admiralty, to propose a route and schedule for a British voyage to Australia – to be led by Matthew Flinders. This proposal was based, explicitly, on his knowledge and suspicions concerning the Baudin expedition.[9] Banks found Baudin's itinerary to be so ill-conceived that he believed it could not be genuine. The real business of the expedition, he opined, was to be carried out at Mauritius and La Réunion – Australia's north-west coast was in his view only the 'alleged destination', intended as an excuse in case the ships encountered British cruisers near the French islands.[10] Rather than worrying about this apparent deception, though, here he saw an opportunity: this 'political manoeuvre' on the part of the Baudin expedition would provide, according to Banks, a crucial advantage to Matthew Flinders.[11] If the French voyage transpired as he expected, Flinders, should he leave in January 1801, would reach the coast of Australia around the same time as Baudin; and, from that point onward, as far as Banks was concerned, the race was on. He suggested that Flinders and his men commence their exploration on the south-west coast, 'in order to secure themselves from being anticipated by the French', and that they make a rough survey 'sufficient to anticipate' the Baudin expedition.[12] Clearly, although Banks had helped to make the French expedition possible, he was determined above all to limit its discoveries and thereby protect British interests. In this sense, while David Mackay argues that strategic aims played only a small part in the voyage of Matthew Flinders, it is clear that in relation to the fate of the Baudin expedition they were in fact significant.

Flinders, himself, was acutely aware that his voyage was intended to forestall Baudin's geographical discoveries. He was clearly reluctant to share navigational knowledge during his 'manifestly unwelcome brush with his rival' at Encounter Bay, as Jean Fornasiero and John West-Sooby explain,[13] and his attitude was no different, later, at Port Jackson. While at anchor there, Flinders wrote to Banks:

> I am happy, Sir Joseph, in announcing to you the success of our voyage this far, and scarcely less so to say that before we met the French ship *Le Géographe* the most interesting part of the south coast of New Holland had undergone the examination of the *Investigator*.[14]

Correspondingly, Banks assured Flinders that according to a French publication he had read, the 'French captains will not be formidable rivals to you, they seem to ... have missed a very many of the openings you have anchored in'.[15] Baudin, for his part, did not show affection or admiration for Flinders but neither do the records suggest that he perceived himself to be in a 'race' with the Englishman. As Fornasiero and West-Sooby remark, he was preoccupied with his own discoveries and, accordingly, he recognized that a cooperative relationship could be beneficial to the work of his expedition. On the south coast, he readily gave Flinders information about his voyage and, upon arriving at Port Jackson, Flinders was one of the first people with whom he sought contact. The two explorers exchanged further navigational knowledge and Baudin obtained from Flinders useful information about the resources of the colony as well as medical treatments. In the end, though, while Baudin chose to remain at anchor for five months so that he could return to the south coast and perfect his investigations there, Flinders returned to sea within ten weeks and sailed directly for the north coast, yet unseen by his French counterpart.

Once Banks had completed the itinerary for Flinders's voyage, he took up his pen again to write to an old friend, Philip Gidly King. Assuming that the *Géographe* and the *Naturaliste* would visit Port Jackson, though it was not included in their itinerary, he instructed King to obtain certain information from the Frenchmen – information which, curiously, did not relate to their discoveries or territorial interests in Australia. He wrote:

> I suspect their principal view is to visit the Isles of France and Bourbon and to come after the inhabitants to prevent them from giving up their allegiance to the Republic ... [I]t will be very desirable that you pick out of any of their people, who will tell you, the history of their visit to the French islands and learn as much as you can of what they have done there.[16]

This was the 'business' that Banks had had in mind when he wrote to Spencer. Clearly, he believed that Bonaparte was at risk of losing Mauritius – one of France's most important but independent colonial possessions. Such a loss to an already diminished Empire would have been a significant blow to the fledgling Republic, particularly during a time of war. Mauritius was also one of the French colonies that most concerned and was most coveted by the British. The behaviour of the colonial administrators toward the Baudin expedition in 1801 demonstrated that they themselves were keenly aware of the desirability of their settlement to their rivals. It was the closest French base to Port Jackson and was perceived as a threat to the general well-being and indeed the security of the British settlement. Throughout the period of the Revolutionary and Napoleonic wars, Mauritius had particularly troubled British mariners and colonists. French men of war and privateers based on the island had been obstructing British ships bringing supplies to Port Jackson and merchants from the French colony had attempted to sell spirits at Port Jackson when King was attempting to curb the liquor trade.[17] Furthermore, King feared that, from their base at Mauritius, the French might make an attack on Port Jackson.[18] If French possession of the island was truly precarious, Banks no doubt considered, it would be ripe for invasion by the British; and, of course, it would be of immense strategic advantage to Britain.[19]

Banks, by presenting the Baudin expedition in this manner, could only have encouraged the governor to perceive his guests and their activities within a political context. In fact, while Banks did mention in his letter that the French ships carried 'men of science of all descriptions', he did not express any interest in those men's work, nor did he instruct King to facilitate their research as would be expected of a leading figure in the commonwealth of learning. Upon first meeting Hamelin, King declared that Banks had sent him a 'pressing recommendation' concerning the expedition. Hamelin referred to this announcement in the context of the warm reception he had received at Government house.[20] If it was the recommendation quoted here, and indeed no other has yet come to light, it certainly was not what Hamelin seems to have imagined.

King's attitude towards the expedition was rather ambivalent: he saw some potential for mutual benefit in providing his assistance to the French but he was also eager, like Banks, to defend the interests of the colony. On the question of the expedition's affairs at Mauritius, his papers remain silent – most likely because there was nothing significant to report – but they do reveal that the visitors had not allayed his concerns about the French colony: King's letter to Lord Hobart about the risk of an attack on Port Jackson was written less than twelve months after the expedition's departure. During the sojourn itself, King was a gracious but cautious host. On one front, he was certainly mindful of the possible intentions of the French government, more so than of the expedition in itself, and of the threats those intentions could pose to the colonial project in New South Wales. On the other, he was concerned about the immediate well-being of the colony in his charge.

Negotiating Authority, Honour and Knowledge

This was not a good time to be receiving a French expedition in the colony: supplies were low – and remember, King was expecting to host the Flinders expedition as well; the number of Irish dissidents, allies of the French and rebellious convicts, had recently swelled; and conflict between the colonists and the local Aboriginal people had very recently come to a head. French notions of liberty and equality were not entirely welcome; freemasons in the colony, for example, had already caused some anxiety for the colony's governors.[21] In fact, just as on the deck of the *Géographe*, ideals about equality needed to be managed carefully at Port Jackson. Consisting of a small and isolated group of settlements, and governed by a naval captain above a large corps of marines, the colonial society was inevitably based upon discipline and hierarchical order. If Baudin's company was more difficult to command when in port, the task of controlling the New South Wales Corps as they hosted 150 of their French counterparts could prove yet more challenging. Indeed, even before the *Géographe* had reached her mooring, King made clear that he intended to supervise the expedition very closely and that, as a consequence, Baudin would face a competing authority at Port Jackson.

The tenor of King's initial directions to the expedition, which were received aboard the *Géographe* as it approached Sydney, left Ronsard feeling deeply insulted on behalf of his commander. The lieutenant observed that King was treating Baudin as his subordinate, despite the fact that the French commander was more highly ranked in the navy than himself:[22] 'I find that he takes the tone of a master and as one would use toward a merchant captain rather than the Commander of an expedition', he declared.[23] Ronsard was a man who, as we have seen, was especially sensitive to the rules of hierarchical etiquette; still, given the broader circumstances, such sensitivity was only to be expected on this occasion. Baudin, for his own part, clearly sensed King's concern and realized, too, that he would need actively to complement the governor's authority. As noted in the previous chapter, when he presented King's regulations to his men, he added a set of his own corresponding but more numerous rules – and thus the standard was set for relations between commander and governor.

Balancing his authority with that of his host was critical for Baudin because, to ensure the well-being of the expedition and to have any chance of preparing a new voyage in this port, he had to be seen by both his men and the British colonists to be firmly in command of the expedition. It was a matter of maintaining enough control to keep the expedition together and even to prepare it for a new scientific voyage. It was also a matter of representing the French, as a pacific, progressive and powerful civilization, before agents of British discovery and empire. Relations between Baudin and King would often reflect a degree of camaraderie, and at times they would bristle with territorial jealousies, but, day to day, they would be shaped mainly by negotiations around Anglo-French politics of knowledge, honour and authority.

At no time during the sojourn were these politics more clear than when Baudin was called to defend the honour of his men. Relations between the French and British officers were at least as fraught as those between Baudin and King – fraught with the same diplomatic issues just highlighted but heated further by resentments and gossip within the New South Wales Corps. The national, martial and masculine honour of each party was called into question on two main occasions – the first, at the commencement of the Republican New Year. To celebrate the occasion, the French officers had dressed ship and, in compliment, officers aboard the British ships alongside them on the harbour followed suit. A misunderstanding soon occurred: the Scottish captain of the *Harrington*, William Campbell, took offence upon noticing that, on the *Géographe* and the *Casuarina*, the British flag was not flying at the head of the mainmast, where it should have been according to the custom of the British navy. He brought it to the attention of the harbourmaster, John Harris, who in turn informed the governor.[24] King ordered Harris to deliver a message to Baudin:

> His excellency the Governor was sorry to say that it had been reported to him that the British flag had been hoisted at the main yard-arm on board the *Géographe* and equally as low on board the *Casuarina*; that being the case, he did not suppose it was done by the knowledge of you, but thro' the mistake of the officers on board, and he was sorry it had not been put in a more conspicuous position.[25]

It was soon found that the French officers had simply been ignorant of the British custom and, as soon as the *faux pas* had been explained to them, they rearranged the flags accordingly. All the same, Baudin took issue with the way in which his hosts had dealt with the issue. He complained to King that 'the thoughtless and careless manner' in which harbourmaster Harris had reported this incident had 'caused, on [Baudin's] part, a letter of reproach and reprimand to his officers who were far from having deserved it'.[26] To Harris himself, Baudin declared that,

> through trusting what you told me, I have sent a bitter and reproachful letter to all of my officers, though it appears from their answers, the truthfulness of which cannot be doubted, that they have scrupulously adhered to the laws of honour, loyalty, and politeness, upon which their conduct is based.[27]

King, assuming that the commander had in fact severely reprimanded his officers, as he had stated in his letter, sent a reply that was intended to placate his guest but which carried a tone that would probably have put Ronsard's teeth on edge:

> I cannot help lamenting your anxiety to show the attention you wished to pay the English nation should have been the cause of your writing the officers on board your

ship a letter of reproach and reprimand for an affair that might have been explained in the same manner as the message was conveyed.[28]

It would appear, based on this exchange, that the commander had overreacted. Yet, there is an inconsistency in this affair. The letter that Baudin wrote to his officers does not in fact accuse them of misconduct, but simply seeks an explanation.[29] Moreover, the officers themselves do not appear to have been affronted. Only Léon Brèvedent, midshipman aboard the *Casuarina*, refers to Baudin's 'reprimand', while Ronsard and Henri Freycinet, who at other times during the sojourn did receive harsh criticism from Baudin and complained about it in no uncertain terms, wrote not a word of protest on this occasion.[30] To King and Harris, then, Baudin deliberately exaggerated the severity with which he had addressed his officers – officers whose honour, he emphasized above all, was beyond reproach.

This was a shrewd diplomatic manoeuvre, based on a genuine sense of injustice that, by scholars depicting his reaction as somewhat unreasonable, is too hastily passed over. Baudin's understanding of and perspective on this episode, explained to Harris, is therefore worth quoting at length:

> If you will glance over the laws of honour of the French Navy, laws to which we have always adhered, you will see at *article 11, chapter 17, page 268, that the place of honour for the flag of a foreign nation which we intend to distinguish must be on the starboard of the main yard arm*. The same law further says: *When it will be necessary to make such distinction, this place will only be occupied by a French flag*. You will therefore see that after having strictly adhered to this rule, I have the right to complain bitterly of the way in which you have proceeded in this matter, as well as the persons who accompanied you, and of the indiscreet tales circulated on the subject. In ignorance of our customs, these tales should at least have been withheld until further information had been obtained. In excuse for yourself and those who complained, you may say that your way of dressing the English vessels is different from ours; but in that case I could answer you that not knowing that way I should never have taken the liberty of passing any remarks, and I could never have imagined that it was out of contempt or other reasons of disrespect that you have not hoisted the French flag at the place assigned by our regulations to that of the nation to which distinction is due.[31]

If 'tales' concerning this incident had indeed been spreading through Sydney to the detriment of the expedition's reputation, and there may well have been given that there certainly were some British officers intent on disrespecting their French visitors, then Baudin's angry protest is not surprising. Principally, it was intended to defend the 'loyalty, honour and politeness' of his officers – those fundamental virtues of the turn-of-the century French gentleman. He also clearly expressed that he alone, and not King, was responsible for their conduct. More broadly, though, Baudin evidently wished to show respect for the customs of his hosts as Forfait had instructed him to do – 'see to it that the religious practices, the political institutions and even the prejudices of the people [who you meet during the voyage] are respected',[32] wrote the Minister – and he felt it was of equal importance that they, in turn, respect French practices and, by extension, the French nation.

The same fundamental issues of national honour and authority were stake in the second incident; though, this time they were not the crux of the problem – instead, they papered over an ongoing colonial dispute. Shortly after the flag affair, Captain Anthony Fenn Kemp falsely accused French officers of selling liquor – in violation of King's ban on liquor trading in the colony and of Baudin's order prohibiting his officers from landing any of the expedition's campaign spirits.[33] Baudin was alarmed at the accusation, but he approached it quite differently than he had done the earlier inquiry regarding the dressing of his ships. Once more, he highlighted the importance of respectful conduct in the circumstances in which they found themselves as well as the questions of honour that were at stake. He declared to King: 'every matter which attacks the honour of an officer is a very delicate one. You well know, also, that suspicion, even if unfounded, is an insult not easily forgiven'.[34]

He also demonstrated his intention to resolve the issue but, this time, while also making a point of aligning his concern and his authority with those of the governor:

> The complaint that was made to you these last days, and which you had the kindness to communicate to me, was of such a nature as to leave you in no doubt that I would do everything in my power to discover to what extent it was founded, and which of those officers who serve in the expedition whose command has been entrusted to me could have dared to disobey your orders and mine in a manner so contrary to the laws of honour of our navy, laws with which you are perfectly well acquainted.

After investigations by King and Baudin, it was concluded that Kemp's allegations had been aimed at discrediting the governor, with a view to restoring the liquor trading rights of the British officers. This, to some extent, explains the much more conciliatory approach that Baudin took to this affair, and the far deeper concern felt by King. To Lieutenant-Governor Paterson, he pointed out that Kemp's actions had caused a

> misunderstanding between the Commodore and the French officers with myself, and every other military officer in the garrison, a misunderstanding which can only tend to do away with the sense they may entertain of the attentions they have hitherto acknowledged, cause an opinion but little honourable to the character of the British officers, and ultimately become the subject of representation between His Majesty and the French Republic, a circumstance that cannot be pleasing to any.[35]

By drawing into the issue the importance of diplomatic relations, King could more easily quash Kemp's underlying protest about the trading rights of British officers. Still, the affair had certainly caused him considerable embarrassment.

Both he and Baudin were motivated to resolve it as quickly and smoothly as possible. In the end, however, the affair was not resolved by them alone. While they corresponded, formally, the French officers sought justice in a more confronting manner, based on principles of honour commonly understood by both themselves and their British counterparts. Employing the power of gossip and scandal, Ronsard and his comrades spread the news of Kemp's dishonest conduct through the town, causing controversy and even frightening the British officer: 'each of us seemed to him to be a spadassin', declared Ronsard.[36] Eventually, Paterson and the officers of the regiment assembled at Kemp's home and, though he resisted, demanded that he make an official apology.[37] Kemp acquiesced, and wrote to Baudin:

> I had no intention of attacking their [the French officers'] honour, that is as far from my purpose as contrary to my opinion of them all. I believe I have proven this by my conduct toward those amongst them with whom I have the honour of being acquainted. My brothers in arms, the officers of the garrison, insist that I tell you that they will not cease to hold the French officers in the esteem that was inspired upon their arrival and throughout their stay in the colony. I am sir your most humble and obedient servant.[38]

The French officers officially pardoned him, but their sense of justice had not been entirely satisfied. Word reached them, possibly through their friend Francis Barrallier, of the New South Wales Corps, that Kemp had 'dared to say that he was angry at having had to apologise and that he would have preferred to fight'.[39] Thus, Baudin's men took further action. Once more, they drew on popular ritual to humiliate the officer socially but, this time, via the production and dissemination of a caricature. They depicted Kemp in uniform with a large padlock on his sword-hilt and wearing an enormous pair of stag's antlers on his head. In the background were two scenes representing the officer's typically conceited behaviour. As Ronsard explained in his journal, Kemp was well known for boasting about his house and his vehicle and, furthermore, his new wife was rumoured to be having an affair with ex-convict lieutenant George Bellasis, with whom Ronsard had developed a friendship.[40] The caricature of Kemp, the boastful cuckold, was circulated throughout Sydney Town and widely appreciated. It was well known even to King[41] and, presumably, to Baudin as well. The records of both governor and commander, however, remained silent on the issue. They were complicit in the Frenchmen's efforts to avenge themselves, and their honour, in this way.

Navigating Territorial Interests

King and Baudin, in fact, while kept busy with the immediate business and affairs of the sojourn, remained mindful during these months of the larger questions hanging over them. The initial meetings between governor and commander, and the French officers as well, had been animated by exchanges of knowledge – discoveries made, coastlines charted, claims laid down – and also of plans for the future. They were exchanges not merely of information, for the purposes of diplomacy and cooperation, but of accomplishments and possessions, performed as patriotic and imperial assertions. Anxieties were unspoken, but, as these exchanges became regular discussions in the subsequent months, they steadily rose toward the surface. In a letter to the Minister of Marine and the Colonies, Baudin noted that 'the English' were in fear of the French establishing a settlement at D'Entrecasteaux Channel. The British government had claimed sovereignty, he explained, over Van Diemen's Land and, he was 'convinced that fearing to have us as neighbours they occupy that part of Van Diemen's Land so as to attempt to authenticate their property right'.[42] Baudin had been open with King about the plans and purposes of his voyage from the time of their initial meeting. He had in fact given the governor access to all of the charts and papers relating to the expedition. Nonetheless, historical circumstances and, raising its head once more, the influence of rumour, conspired to fuel King's fears.

Shortly after the French ships farewelled the port and set sail for the south coast, the governor sent a party aboard the *Cumberland* ostensibly to establish a British presence at D'Entrecasteaux Channel but, in effect, generally to keep watch on them. There may have been no suggestion of French territorial plans from Baudin, during the Port Jackson sojourn, but it seems there had been from the young and ambitious François Péron. And, once the Frenchmen were actually out of the colony and heading toward Tasmania, these suggestions sparked an explosion of rumour and speculation throughout Sydney Town. Baudin met the *Cumberland* at King Island, and without delay, he wrote to the governor. He made it clear that the rumours had been false, but added that,

> in any case, you must have been quite sure that, if the French government had ordered me to stop for several days in the north or south of Van Diemen's Land, discovered by Abel Tasman, I would have done so, and without keeping it a secret from you.[43]

He chided King for his distrustfulness, but this was also a final jostle for personal authority, as well as for national sovereignty. Baudin questioned the British government's rights to claim possession of Tasmania – it was not the British who had discovered this island, he pointed out, and neither was it the British who had explored it most thoroughly or charted it most accurately:

if it were sufficient, according to the principle you have adopted, to have completely explored a country for it to belong to the person who first made it known, you would have no claim at all. To convince yourself that it is not the English, you have only to cast an eye over the idealised maps drawn up by your geographer Arrowsmith and compare them with the ones done by Beautemps-Beaupré, which leave very little to be desired.[44]

This argument was also an assertion of the superiority of French knowledge. The British could take physical possession of Tasmania, Baudin implied proudly, but that would not change the fact that, intellectually, the strongest claim was held by the French.

Baudin in fact wrote two letters to the governor from King Island – one addressed to King as 'the Governor General of the English settlements at New South Wales', much the briefer of the two, and the other 'to my friend Mr King', an open and detailed account of how the author viewed the colonial project. Much has been made of the friendship between King and Baudin, and there is a degree of candour in this second letter that points to a genuine feeling of intimacy. However, while it is the nature of friendship that such intimacy must have been reciprocal, it is worth noting that King's correspondence concerning Baudin reveals markedly less warmth and more circumspection than do Baudin's letters to him. Opening himself up to friendship with an English colonial governor had its risks but was not highly problematic for the French commander – a guest in the region, seeking cooperation and, with any luck, companionship as well. King, for his part, could not forget that Baudin represented a potential threat to the colonial project and, correspondingly, to his political career. As Penny Russell shows, the nascent colony was a space of social and geographical mobility. The 'race for fortune', for advancement or even for survival, often a race to return 'home', preoccupied all its new inhabitants and none more so than King.[45] His future depended not only on individual success but on the prosperity and expansion of the colony. For him, it was not entirely the case that 'all friendship is form, all confidence mere empty show, and valued accordingly',[46] as one later visitor to Sydney would remark of the local culture, but his attitude toward the French commander was certainly composed largely of manners, suspicion and diplomacy. He kept a souvenir of Baudin's visit: a portrait sketched at his request. He also wrote for Baudin a letter of 'introduction to [the] protection and good offices' of Sir Joseph Banks, stating that he had 'taken the liberty of introducing a worthy and respectable officer to [Banks's] list of scientific and literary acquaintances'.[47] Yet, in 1803, King remarked to Lord Hobart that he deemed the establishment of a British settlement at Van Diemen's Land 'the more essential from the inclination the French have shewn to keep up a correspondence here'. There is not only mistrust represented in this statement, but also a willingness to use his relationship with Baudin to hasten the expansion of the colony – that is, to play on the fears in London of French rivalry. In any case, the governor was no doubt Baudin's 'friend Mr King', yet his role as 'Governor General of the English settlements at New South Wales' would always predominate.

It would be naive to assume that it was only on King's side of the friendship that ulterior considerations may be found. By befriending the governor, Baudin's chances of obtaining the cooperation essential to the continuance, indeed renewal, of his voyage were considerably improved. And, though there is no reason to doubt the sincerity of his wish to maintain the friendship beyond the period of the sojourn, the governor may have had a point: it is entirely possible that Baudin also thought to remain informed of British activities in New South Wales. Already, the months he had spent in Sydney had enabled him to send detailed information to the Minister of Marine about the state of the colony and about King's intention to settle Tasmania. Baudin was not blind to the strategic and even territorial gains that his discoveries during the course of the voyage could afford the French government. When informing the Minister of the prospect of a British colony in Tasmania, he remarked: 'it will be a real loss for France, for an establishment in the south of Van Diemen's Land can only procure great advantages for commerce and seems to be what politics dictate'. And later, this time from Timor, he implied to the Minister that he held information which ought to be guarded:

> I am not entering into further details as this letter will perhaps pass through several hands before reaching you, and it would be malapropos for it not to reach you intact, given that I have something interesting to communicate to you.[48]

It is quite likely that the 'something interesting' was information about the British colonial project.

It is well known that certain members of Baudin's expedition had in fact thought about the potential for the French not only to rival the British colony in Australia but to invade it. Hamelin, as shown earlier, had dared to put such thoughts on paper while still in Sydney Harbour.[49] His first lieutenant, Pierre Milius and naturalist François Péron, for their own parts, each wrote a report during their homeward stopover at Mauritius – at the request, it would seem, of the governor of the French colony, General Decaen.[50] As noted earlier, Mauritius had long played an important role in Anglo-French imperial politics, but, by the time the *Géographe* called in to Port Louis in August 1803, the Napoleonic Wars had commenced and Decaen felt the colony was particularly vulnerable. He wished to persuade Bonaparte of the need to weaken Britain's influence in the southern seas via an attack on India – and, he proposed more hesitantly, perhaps on Port Jackson as well.

The First Consul, however, had set his sights elsewhere. He was eager to defeat the British but to do so closer to home; preferably, by attacking Britain itself. Decaen's plan did not eventuate and, moreover, although versions of at least Péron's report were distributed in Paris – to Fourcroy, director of the Muséum d'Histoire Naturelle and to Fleurieu, interim naval minister[51] – French authorities gave little response to any of the strategic information on Port Jackson from Baudin's expedition. It is not surprising that some of Baudin's men looked at Port Jackson with a martial eye, but it has been well established that they did not do so under instruction. In fact, this expedition of the 'Napoleonic era' was less political than the earlier expeditions of the Enlightenment. La Pérouse, for example, had been issued specific and detailed instructions on 'subjects relating to politics and commerce'. Louis XVI had ordered:

> In all the islands, and harbours of the continent, occupied or frequented by Europeans, at which he shall touch he will consider it as a general rule, to make with prudence, and as far as circumstances and the length of his stay shall allow, every inquiry, that can enable him to ascertain with some minuteness, the nature and extent of the trade with every nation, the naval and military force which they maintain there.[52]

Moreover, Port Jackson had not been included in Baudin's itinerary and there is no persuasive reason why an intended reconnaissance of the colony would have been kept secret. Even Banks had not proposed such a notion.

In regard to the British in Australia, Baudin's expedition was at most a symbol of French rivalry. It was largely for that reason that relations between the colonists and the voyagers were cooperative. They were cooperative in part, too, because of the tradition of scientific internationalism. However, they were not apart from the Anglo-French tensions that existed either historically in the Pacific or contemporaneously in Europe, but in fact clearly interpreted them in this place and moment. Indeed, these tensions both caused and helped to resolve discord throughout the sojourn and even beyond, to the expedition's stopover at King Island. King and Baudin needed to cooperate in order to avoid sparking further hostilities between their nations as well as, on the one hand, to defend the interests of the colony and, on the other, to protect the well-being of the expedition. And they did this not only through the provision of accommodation and of access to official papers and information. They drew on a shared understanding of Anglo-French history in the Pacific – a history of exchange and mutual benefit as well as of rivalry – of standards of naval and manly honour, of popular and official cultural rituals, and of the importance of naval tradition. Still, it was the governor who was in the position of most power and who had most reason to feel defensive. There was naturally as much political prudence and good manners in his hospitality as there was warmth and generosity. Between King and the French commander, relations of authority and respect were being constantly negotiated. Their association was diplomatic, above all, and Baudin was not a passive guest, dependent entirely upon the kindness of his host – if he had been, he would, at best, have been able to prepare only for a homeward voyage.

5 LIBERTY, EQUALITY AND 'CIVILIZATION': OBSERVATIONS OF COLONIAL ABORIGINES

In the letter to his 'friend Mr King', Baudin in fact wrote less about British claims to the possession of Tasmania than about the colonization of Indigenous Australians. On this point, he was again concerned about territory, though not in relation to the rights of discovery but to the laws of nature and the corresponding duties of government. 'I have never been able to conceive', revealed the commander,

> that there was any justice or even loyalty on the part of Europeans in seizing, in the name of the government, a land they have seen for the first time when it is inhabited by men who did not always deserve the titles of 'savages' and 'cannibals' that have been lavished on them, whereas, they were still only nature's children and no more uncivilised than your Scottish Highlanders of today or our peasants of Lower Brittany, etc., who, if they do not eat their fellow men, are no less harmful to them for all that.

Baudin opined, in fact, that

> it would be infinitely more glorious for your nation as for my own to train for society the inhabitants of the countries over which they each have rights, rather than undertaking to educate those who live far away by first seizing the land that belongs to them and to which they belong by birth.

Based on a firm assumption about the universality of the human race, it was both the rights and the progress of humanity – across the divides of class and region as well as colour – that concerned Baudin. These were the concerns that moved him to criticize the British colonial project in New South Wales. If the above 'principle had generally been adopted' by King and his fellow British authorities, argued Baudin, they 'would not have had to form a colony of men branded by the law and abandoned to themselves. It thus follows', he continued

> That not only do you have an injustice to reproach yourself with, in seizing the land, but you have also transported to a land where the crimes and diseases of Europeans

were unknown everything that could retard the progress of civilisation, and that was
used as a pretext by your government, etc.

Baudin had in fact closely observed how the penal colony had affected the local
Aboriginal people. He explained to King in forthright terms:

> If you would reflect on the conduct of the natives since you first settled on their ter-
> ritory, you will see that the distance they keep from you and from your customs was
> brought about by the idea they formed of the men who wished to live with them. In
> spite of the precautions and the punishments you dealt out to those who mistreated
> them, they were able to discern your future projects, but, being too weak to resist you,
> the fear of your weapons has made them leave their land, so that the hope of seeing
> them mix among you is lost, and you will soon be left the peaceful possessors of their
> birthright, as the small number of them living around you will not last long.[1]

Five months ashore in a penal colony, five months reassessing the objectives and
rationale of the expedition, were bound to bring social anxieties, beliefs about
human rights and ideas of 'civilization' to the fore. This colonial encounter
was therefore bound, too, to produce particularly revealing representations of
Aboriginal people – revealing, that is, in regard to the *mentalités* of the French
observers, if not to the culture and characteristics of the locals themselves.

The Port Jackson sojourn provided Baudin and his men with a unique
opportunity to extend their ethnographical work. This was the voyagers' only
opportunity to observe how Aboriginal people responded to ongoing contact
with European society and, more precisely, how they responded to European
efforts to 'civilize' them. It was also by far their most prolonged period of contact
with 'the natives' of Australia. Not surprisingly, their records of this encounter
evince different attitudes toward ethnographic study, toward Aboriginal people
themselves, and at times toward European society as well, than those expressed
at any other point in the course of the voyage. However, while one finds in them
small patches of detail, and sometimes glimpses of insight, they overall evince
a sense of detachment. They comprise mainly summaries and evaluations of
characteristics widely observed, rather than recounts of actual contact, of the
gestures, words or gifts exchanged with individuals. And, on the situation of the
Aborigines within and around the space of the colony, they feature significant
silences, a circumstance on which scholarship has remained similarly quiet.

The sparseness of the textual ethnographies is most often attributed to a
putative view, on the part of the Frenchmen, that the Aboriginal people of Port
Jackson were no longer pristine examples of Indigenous humanity and thus not
suitable ethnographic subjects.[2] Certainly, in a treatise written in 1800 and given
to the Baudin expedition to guide its observations in anthropology,[3] *ideologue*
Joseph-Marie Degérando had identified just two human conditions: 'civilized'
and thus 'modified by a thousand various circumstances, by education, climate,

political institutions'; and 'savage', affected only by 'natural', 'primary and fundamental' circumstances, which 'belong to the very principle of existence'.[4] He made no mention of 'savages' who were in contact with 'civilizing' influences and had adapted their practices and behaviours as necessary but had not conformed to 'civilized' society.[5]

However, this point of view was not specific to its time and it did not pose such an obstacle to either previous or subsequent voyager-ethnographers in the colony. For example, while visiting Port Jackson in 1793 Alessandro Malaspina noted that observation of the local Aboriginal people was indeed problematic: there was a risk of 'confusing the Indigenous customs with those which have been imperfectly adapted from Europeans', he explained. Yet that risk did not stop him. Malaspina went on all the same to describe the Aboriginal people and how they were dealing with colonization.[6] Furthermore, following his visit in 1820, Louis Freycinet published a comprehensive account of Aboriginal people in their colonial context: how the settlers had tried to 'civilize' them, the degree to which these attempts had worked and how Aboriginal people had often refused to conform, as well as how Indigenous inhabitants had been affected by dispossession and the introduction of alcohol to the region.[7] Clearly, it was particularly to the views held by Baudin and his men that this colonial scenario posed a profound challenge.

Approaching the 'Other', Within and Without

As explained earlier, the Baudin expedition had been constructed during a key moment in the natural history of man. It was an ideological and especially a methodological turning point, produced by the transition from the Enlightenment pursuit of encyclopaedic knowledge to disciplinary specialization, impelled by a sense of urgency among French naturalists to reach a more profound understanding of human nature, and, in all, arising out of the upheaval and principles of the French Revolution.[8] In fact, by the turn of the century, 'liberty, equality and fraternity' had at least theoretically been granted to the French people, slavery in the colonies had for the time being been abolished and the new Republic, embracing all the citizens of France, had been proclaimed 'une et indivisible'; as Martin Staum notes, it was time to 'stabilize the Revolution'.[9] This entailed investigating how well the Republic's democratic principles were actually suited to the 'laws of nature'.[10] Thus it was that, while naturalists pursued their investigations with increasing vigour, under instruction from the Directory and then the Consulate, prefects across France embarked on comprehensive statistical surveys of the people under their authority. Moreover, for the amateur ethnographers in the countryside and the naturalists in the Muséum, the 'voice of reason no longer sufficed'.[11] With precise, objective and comparative methods,

Man was to be studied much as were plants or minerals: as a product of and in its relationship to nature.

The result of such a thorough approach to the regional surveys was the discovery of greater diversity, and a greater lack of social and ideological progress among the rural communities of France than officials had expected to find – 'un monde sauvage',[12] in effect. The Directory and the Consulate in turn sought to erase this savage world, with its diversity of patois and its enduring superstitions. Those people whose 'negative' characteristics naturalists could not explain by reference to environmental influences rather than to human nature itself were put aside as examples of 'the sordid and ugly side of the *France imaginaire*'.[13] Faith in the unity of the human race had become so vital to the French, asserts Bourguet, that the concept of equality had begun to merge with their sense of national identity,[14] and Revolutionary naturalists looked essentially for human similarity.[15] Many, such as Louis-François Jauffret, secretary of the Société des Observateurs de l'Homme, sought also to renew faith in the *grandeur de l'homme*. In 1800, Jauffret declared:

> The Société, in seeking to raise human dignity, this beautiful prerogative that was so cruelly misunderstood, so insolently outraged during the dreadful regime that weighed down France awhile, will have the advantage to contribute, just by the influence of its observations, to the eradication of a mass of abuses that that odious regime introduced, and that the current government has not yet been able to destroy completely.[16]

In line with the development of a more 'scientific' approach to the study of man,[17] the purpose of the Société was to gather comprehensive facts and observations – 'leaving aside ... vain theories' and 'hazardous speculations' – on the moral, intellectual and physical aspects of Man.[18] In particular, it intended to draw on anatomy, physiology, medicine and hygiene – a 'particular direction that would offer newer and more important research'.[19] This all-encompassing and ambitious approach was representative not only of the rapid development of the natural history of Man, nor of the great importance placed on the attainment of a deeper understanding of society. It also represented a sense of optimism – optimism that adequate research would explain away the differences between peoples, confirm a solid thread of similarity and ultimately pave the way to social and political progress.

This moment as a whole was reflected in the very nature of Baudin's expedition. Amongst the staff of twenty-two naturalists and artists was medical student François Péron. After presenting to the Institut a paper entitled 'Observations sur l'Anthropologie', Péron established himself as the expedition's 'observer of man',[20] with the aim of testing his hypothesis that civilization was detrimental to human health. His continued focus on the physical during the course of the expedition ultimately marked a turn toward the nineteenth-century 'science of Man'. So too did the fact that in addition to Louis XVI's 'Of the Conduct to be Observed Toward the Natives',[21] given earlier to La Pérouse, this expedition carried specialist instructions concerning the observation of Indigenous peoples. As Margaret Sankey states, these documents constituted the first codification of 'anthropological' methods.[22] At the request of the Société des Observateurs de l'Homme, Baudin was provided with two documents that represented distinctly divergent approaches to ethnographic research. One document was the treatise composed by Degérando, which concerned comprehensive research about savage society and culture and emphasized the concept that human diversity reflected only the 'ages of human society'. The other was a set of instructions from comparative anatomist Georges Cuvier. This was designed to guide the voyagers in the collection of anatomical specimens and data that might advance classification of the 'races' of humankind.[23] Despite their differences, however, both guides highlighted the importance of a more systematic, thorough and objective approach than that taken in the past.

Confronting Expectations

In practice, Baudin and his men did not closely adhere to their instructions, though that was largely because they did not have the opportunity either to immerse themselves for a prolonged period in 'native' society as Degérando advised, or to gather the quantity of skeletal remains that Cuvier desired; however, their various methods did reflect the transitional nature of anthropology at this time and they certainly carried an optimistic humanism to the shores of Australia. The expedition's first encounter with Aboriginal people occurred at Geographe Bay in 1801. While the voyagers found little to admire in these natives, their disappointment related more to what had in fact been 'non-encounters'[24] – the Aborigines had signalled that the visitors should leave before fleeing from their approach – than to the people themselves. The voyagers' faith in the 'good savage' philosophy endured well into their encounters in Tasmania almost eight months later. Baudin and his men arrived at d'Entrecasteaux Channel with little doubt they were to meet the 'good' and 'peaceful' Tasmanians enthusiastically described by Jacques-Julien de Labilliardiere following his visit to this region with Bruni d'Entrecasteaux in 1792.[25]

However, by the time they departed Tasmanian shores for Bass Strait, they were feeling disillusioned. They had been offended by what seemed to them ingratitude or brutishness on the part of the Aboriginal people: their gifts had been rejected, the artist's refusal to hand over a portrait had caused anger and, twice, attempts at kindness and generosity had been returned with violence.[26] Moreover, the Tasmanians' bodies had not matched their expectations either.

François Péron, who was particularly interested in this point, claimed that both the physical strength and the sexual libido of these people were considerably weaker than those of Europeans. He posited that this was a result of the adversities of the Tasmanians' 'natural state'.[27]

Yet it was not the Aboriginal people themselves who disappointed and frustrated the voyagers most of all but the French savants who had idealized them, alongside other Indigenous peoples. In this way, their attitude strongly reflected the naturalist and national turn that had affected the study of Man since d'Entrecasteaux's expedition. Baudin, for example, sternly reproached Labillardière, who he points out had been 'employed only as botanist', for his 'fertile imagination',[28] while Péron, especially incensed, declared that the Tasmanian 'is pre-eminently, *the child of Nature*', yet 'how greatly he differs ... in intellect and physique from those alluring images of him that fancy and enthusiasm created and that stupid obstinacy then wanted to set up against our social state!'[29] It seemed that the theory Péron had presented to the Institut had been thoroughly disproved and, in fact, in the official account of the voyage he would argue the reverse: that social, cultural and political development (in the European sense) improves the human condition.[30]

His and his fellow voyagers' assumptions had been unsettled by a comparatively objective approach. They had all been led to regard the Tasmanians in a more critical light than their predecessor had done. Yet, they still understood this example of the human race not as innately flawed but as limited in its development by environmental factors and thus child-like. At this stage of the voyage, the voyagers did not question the potential of Tasmanians or of Aboriginal people in general eventually to be 'civilized'.

Encountering the Colonization

Upon arriving at Port Jackson, Baudin and his men were therefore confident they would witness this potential being reached, for amongst the great variety of insights, biases and details to be found in their records of the Indigenous inhabitants, there is one distinct repetition: colonization had not in fact civilized them. Baudin reported that the only progress they had made was in making 'more progress in the English language than the English have in theirs',[31] while with a sense of astonishment botanist Théodore Leschenault de la Tour commented: 'although, for several years, the natives of the environs of Sydney have been visiting the English ceaselessly and without fear, they are nonetheless hardly less barbaric than before the arrival of Europeans'.[32] Lieutenant Pierre Milius observed the same, while Péron declared: 'they still live amidst war and alarm'.[33] This preoccupation was symptomatic of the particular difficulty facing the observers of the Baudin expedition. Unlike Malaspina before them, their challenge was not merely describing Aboriginal people who were no longer pristine, but explaining why these people were not advancing along the path to civilization and what this meant about Man's relation to nature and its capacity for progress.

In actual fact, there was more to the Aborigines' situation than a lack of 'progress', and it is only Baudin's letter to King – ironically, as that document was in no way intended as an ethnography – that alludes to another side of the story. The commander focused not on the nature of the Aboriginal people themselves but on their contact with the British colonists and convicts. He referred to the detrimental influence of the 'crimes and diseases' introduced to Australia through the penal colony, a spirit of active resistance to colonialization – if ultimately he deemed the Aborigines 'too weak to resist' and, hinting that there had been violence on the part of the colonists, he also mentioned the Aborigines's 'fear of ... weapons'. He asserted that it was not merely due to the Aboriginal peoples' nature but also because of the conduct of the British that this project of civilization was failing.

Baudin took his observations on the cross-cultural conflict so far, but neither in this letter nor in his letter to Jussieu at the Muséum did he discuss what contemporary historians describe as a colonial war – and one that had reached a climax just shortly before the Frenchmen's sojourn. Around twelve months before the arrival of the *Naturaliste*, Governor King had reacted to an Aboriginal campaign to drive out the colonists by ordering settlers to use gunfire to force all Aborigines away from colonial residences. Six months later, he outlawed the campaign's leader, Pemulwuy, himself.[34] Pemulwuy was killed sometime in the first half of 1802 – possibly during or after the *Naturaliste*'s first visit from late April until early May, but before the longer visit of both ships from late June. The Frenchmen must even have known the man responsible: Henry Hacking joined the Flinders expedition during its stopover at Port Jackson.[35] Following Pemulwuy's death, notes Grace Karskens, a 'second coming in' occurred;[36] still, continuing tensions must have been palpable throughout the colony. In fact, desperate to bring peace, King had not only offered a reward to colonists for killing Pemulwuy but also a promise to the local Aborigines: the war would end if they gave up the rebel leader.[37] His guests could not have missed the consequent hostility between Aboriginal groups, between Black and White, and between the colonists as well – those who resisted King's orders, such as the naturalist George Caley and those keen to support it, such as Samuel Marsden.[38]

Grace Karskens notes that the English themselves were vague about the details of Pemulwuy's death.[39] Perhaps they felt some shame about the lengths they had gone to in order to quell the conflict or, more precisely, about what these actions said of their humanism and authority. The voyagers, however, need not have perceived the situation from that angle: even though this sojourn was unplanned, as scientific voyagers at the turn of the nineteenth century, their objective was to contribute to the natural history of Man. Indeed, it is tempting to suggest that what disturbed these observors above all was not merely the fact that the Aboriginal people had not yet been assimilated into this portion of British society, that they were not, for instance, taking up domestic service on an ongoing basis or either dressing like Europeans or living in European-style housing – as shall be shown, some mention was made of this sort of lack of 'improvement'. Perhaps what these post-Revolutionary observers found most confronting of all and most difficult to explain was that circumstance omitted from all their records: a sustained attempt to 'civilize' failing to the point of violence and perhaps even, as Baudin predicted, to the point that the local people would 'not last long'.

If the Frenchmen were dispirited by this scene, then their disappointment is likely only to have been exacerbated by the relatively uninspiring and unsettling nature of the encounter itself. To begin with, what could the voyagers hope to bring to these people? Degérando had wished French voyager-naturalists to offer Indigenous people 'the pact of a fraternal alliance!' and to 'take their hand and raise them to a happier state: 'Bring them our arts, and not our corruption, the code of our morality, and not the example of our vices, our sciences, and not our scepticism, the advantages of civilization, and not its abuses.'[40]

The Aborigines of Port Jackson had already been introduced to the 'abuses' of 'civilization' and the corruption and vices of 'civilized' people. A relationship had already been established, harm already done. The Frenchmen were thus deprived of the gratification of contributing to a civilizing effort and, at least for Baudin, also of pride in being associated as fellow Europeans with the colonists' endeavour. It is significant that the colonial space deprived them furthermore of the exhilarating challenge of making contact and attempting exchanges. These Aboriginal people were habituated to Europeans, their ways and their products. Moreover, as Baudin noted, they even spoke some English. In fact, the encounter was therefore not only less engaging than their previous beach encounters but, as it did not call for the usual routines and gestures, it forced the voyagers to relate to the Aborigines in ways they had not done before. Moreover, in contrast to meetings on Indigenous beaches, it forced them to relate to these people on a day-to-day basis, on the streets of a European settlement, over the course of five months. In all, this encounter would confront the Frenchmen with ordinary Aboriginal humanity set starkly against the Europeanized backdrop. Positive 'noble savage' and 'child of nature' paradigms were fed by a distance created by only fleeting contact, communication barriers and novelty; two seasons in a British colony were not likely to provide that same space, but they would lead the French observers themselves to distance the Aborigines' humanity in negative ways from their own.[41]

Grappling with Difference

It is already evident that during this sojourn Baudin, at least, had not tended to look detachedly at an 'Other' but rather had sought to understand the Aboriginal people with consideration to the various lives and roles within a close and universal human society. For him, the colonial contact zone evoked thoughts of the social issues at home and rendered the humanity of the Aboriginal inhabitants, in a sense, more familiar. When he likened Port Jackson's Aboriginal people to the Bretons of France, the commander demonstrated a concept of humanity that focused on similarity instead of difference. At the same time, he associated the Aborigines with a cultural stereotype well known to the French – as Bourguet comments, the Breton world, in all, was the negative example of France *par excellence*, of 'the world of dearth, of wasteland, a world where bad herbs and superstition proliferate'.[42] His suggestion that the British should concern themselves with their own nation rather than with the Aboriginal people of Australia thus reflects not only his belief in human rights but also his concern about European civilization itself – heightened by the sight of criminal society transplanted onto the 'pristine' environment of Port Jackson and no doubt too by a sharpening of his identity as a Frenchman.

His attitude also indicates a belief that Aboriginal people had not the capacity or the strength to 'last long' in the face of European colonization. This was the first time in the course of the voyage that Baudin had observed in Aboriginal society, or suggested about human nature more generally, a vulnerability, weakness or potential incapacity to adapt to foreign ways. His view of Aborigines in the colonial environment led him to write more pessimistically about the nature of Man, in its broadest sense, than he had done on the shores of western Australia or Tasmania. He demonstrated loss of hope in a people's ability to survive let alone to progress in a situation of adversity; though where lay the primary cause, he did not speculate.

Certain of his men, however, without going as far as invoking what seems to have been the emergent 'dying race' theory, did not hesitate to theorize about the perceived limitations of Port Jackson's Aborigines. Leschenault and Milius were vague about the particular circumstances of colonization at Port Jackson. They joined their commander in highlighting that 'civilization has made no progress among these people in the 15 years the English have inhabited this island',[43] yet made no mention of the colonial war or of Pemulwuy's execution. Leschenault avoided any reference to how the Aboriginal people actually lived and responded to Europeans within the colonial space. Still, though by distinctly divergent reasoning, they both concluded that Port Jackson's Aboriginal people would never advance to a more 'civilized' state.

Leschenault posited a theory centred on his belief in the natural environment's predominant influence over human development. He suggested that the 'the pressing need to defend their existence' had 'destroyed ... whatever happy moral and intellectual qualities' with which one might credit these 'natives'. Furthermore, he continued: 'nature appears to have endowed them with just the sum of intelligence in harmony with the land they inhabit'. In fact, in Leschenault's point of view, nature had done its job all too well: he saw scant hope that the Aborigines' natural 'sum of intelligence' would develop any further for 'never, at any time that we had occassion to communicate with them', he declared, 'did we notice that degree of curiosity that indicates aptitude and desire for learning'.[44]

As for Milius, he wrote not of any actual incapacity on the part of Aboriginal people to 'change condition' but rather – and here he refers particularly to male Aborigines – of a lack of 'desire' and a 'natural penchant' for 'indolence'.[45] This claim he attempts to substantiate with specific examples taken from the colonial context. Firstly, he refers to an anecdote, heard from Governor King, about individuals who ran away from domestic service to resume their 'indolent' ways – which supposedly involved depending on the forced labour of Aboriginal women.[46] Secondly, he wrote specifically of Bennelong, who some years earlier had visited London with Arthur Phillip, and of this man's obvious 'repugnance' for European ways. According to the lieutenant, it was ultimately that negative feeling which showed it was 'impossible to expect to bring the savages of these lands to any idea of civilization'. They were, he concluded, 'truly stupid brutes who must be left to live their own way'.[47] This vague notion of Indigenous men's laziness, ignorance and brutality had, as Shino Konishi demonstrates, a long history in Enlightenment thought and exploration narratives;[48] thus, it is interesting that Milius was the only member of the expedition to engage with it so directly, to draw on the words of individuals such as King and Bennelong to set it in a context bound to interest European readers, and to use it support a claim that Aboriginal people would not be civilized.

Certainly, Péron's approach to the question of the Aboriginal peoples' lack of 'progress' was markedly different. At Port Jackson, he continued to make observations about the moral, intellectual and, in particular, the physical attributes of the people he encountered. He used a dynamometer to test the strength of his subjects and employed the results of these tests above all to substantiate his theory about 'civilized man's' superiority over the 'savage'.[49] Indeed, in contrast to Baudin, his contact with the British colonial project seemed to bolster his faith in the importance of colonization in regard to the 'equalizing' of human societies. He even extolled the virtues of the transportation system: he applauded the improved morality and fertility of the convicts and attributed these improvements to the healthy climate, fresh air, varied diet, as well as to the orderly social organization established by the colonial authorities.[50] However, the local Aboriginal people were relegated to the very margins of his voyage narrative and conclusions about them were presented only indirectly in a separate ethnographical chapter. This work constituted a synthesis of the ethnographic observations Péron had made throughout the duration of the voyage and of his subsequent textual research. It was a carefully considered presentation of Péron's theory on the benefits of civilization.[51]

Here was a strong reflection of the changing nature of French approaches to studying man – a change that was not to bode well for colonized peoples.[52] In Péron's concentration on the comparative analysis of anthropometric statistics, as well as in his claim that Tasmanians and the inhabitants of New Holland were of different origins,[53] one sees early signs of racialist thinking not uncommon at this time in the physical disciplines: comparative anatomy, physiology and medicine.[54] Péron's approach was to rank the peoples encountered during the voyage according to calculations of their physical strength and assessment of their level of social development. The Indigenous people of 'New Holland' were ranked above those of Tasmania but below the peoples of Timor, France and Britain.[55] According to Péron, the differences between all these groups had been caused by the advantages or disadvantages of their respective environments, particularly their diets and dwellings.[56] Interestingly, however, Péron's published report does not mention how the people of 'New-Holland' had responded to the presence of the British society. Much less does it give an explanation as to why, as Péron asserted in an unpublished report, they had not changed since the establishment of the British settlement. While excluding the colony from view and distancing the Aborigines' humanity from that of Europeans, through references to their 'fierce, vagabond ways',[57] Péron represented the mainland Aborigines as 'savages' – untouched, if not 'good'. More specifically, he described them as 'deprived children of Nature' rendered weak and left to 'vegetate' in the savage state by, as Leschenault also argued, the 'lack of food, its poor quality, and the labour needed to obtain it.'[58] Yet, the condition of these 'miserable people' was not fixed, he claimed. Demonstrating faith in a gradual and organic process of human civilization – perhaps an alternative to the value of European intervention which, like his companions, he seemed unable to illustrate – Péron imagined the Aboriginal people eventually forming villages, farming kangaroos, growing stronger, smarter and more morally refined.[59]

Péron's vision of Aboriginal civilization feels rather forced, despite, and indeed partly due to, his own claim to the contrary.[60] On this point, it is worth noting that Péron was a protegé of Georges Cuvier, and also to recall that since his presentation to the Institut in 1800 it had been his aim to make a significant contribution to the nascent field of anthropology. By the time he reached Port Jackson he had had to reverse his initial hypothesis: it was now centred on the profound benefits of civilization. Therefore, while Péron could afford to make the daring claim that human groups might be divided by different origins, to decry the Aborigines' existing condition, and even to assert that these people had until that point failed to embrace civilizing influences, it would have been detrimental to his argument to posit, like certain of his colleagues did, that the colonization of Aboriginal people was doomed to failure.

Some of Baudin's men preferred not to take up the issue of colonization at all in their ethnographic observations. Not only were there no references to the colonial context in their ethnographic descriptions but there was also no speculation about the Aborigines' civilizing potential. Geographer Charles-Pierre Boullanger and artist Charles-Alexandre Lesueur described Aboriginal practices such as fishing, cooking and childbirth.[61] Lesueur also depicted certain of these practices in a series of 'typical landscapes' (see Figure 5.1 on p. 93 and Figure 5.2 on p. 94).[62] Doubtless, these detailed records would have been of value to the savants at the Muséum in Paris; however, there is no sign that they were seen there and scant indication that Péron drew on the reports when he prepared the official ethnography for publication, though he did include the landscapes in the *Atlas* of the *Voyage de découvertes*. This apparent neglect of Lesueur and Boullanger's observations and the determination of these observers themselves to detach their Aboriginal subjects from the colonial context further highlight the degree to which the Frenchmen's anthropological gaze was problematized by the issues of colonization.

Granted, there is no doubt that the circumstances of this encounter produced more empathetic visual records of Aboriginal people than those produced earlier in the course of the voyage. The portraits by Petit are particularly notable for their signs that the artist recognized his subjects' common humanity. In Tasmania, anxious and relatively distant relations with the Aboriginal people had led Petit to produce somewhat wild, even caricatured, representations of his subjects.[63] By contrast, at Port Jackson, where local Aborigines were familiar with European contact and sometimes with being sketched, Petit was able to enjoy sustained contact with his subjects and come to know them better, and produced far more polished and lifelike portraits (see Figures 5.3, 5.4 and 5.5 on pp. 95–7).[64] Further, while Petit generally abided by the instructions on portraiture provided by Georges Cuvier – instructions intended to influence portraits that could advance Cuvier's research linking cranial dimension, facial angle and physiognomy to moral and intellectual features, at Port Jackson, he represented individuals rather than merely 'racial' types.[65]

NOUVELLE – HOLLANDE : *NOUVELLE-GALLES DU SUD.*

Grottes, chasse et pêche des sauvages du Port-Jackson.

Figure 5.1: 'Grottes, chasse et pêche des sauvages du port Jackson, à la Nouvelle-Holland'; engraving by A. Delvaux after C.-A. Lesueur, published in the *Atlas* of the *Voyage de découvertes aux terres australes* (1824), plate 31. Source: National Library of Australia, an7569797.

NOUVELLE - HOLLANDE : N.^{lle} GALLES DU SUD.

NAVIGATION.

Figure 5.2: 'Navigation', engraving by C.-F. Fortier after C.-A. Lesueur, published in the *Atlas* of the *Voyage de découvertes aux terres australes* (1824), plate 34. Source: National Library of Australia, PIC/11195/34 NK1429 LOC NL shelves 599 (Atlas).

NOUVELLE-HOLLANDE : N⁰ᶩᵉ GALLES DU SUD.

OUROU-MARÉ, dit BULL-DOG par les Anglais, Jeune guerrier de la Tribu des GWÉA-GAL.

Figure 5.3: 'Ourou-maré (dit *Bull-dog*), jeune guerrier de la tribu des Gwea-gal', engraving by B. Roger after N.-M. Petit, published in the *Atlas* of the *Voyage de découvertes aux terres australes* (1824), plate 23. Source: National Library of Australia, an7569776.

Figure 5.4: 'Norou-gal-derri, guerrier des environs de port Jackson, s'avançant pour combattre', engraving by B. Roger after N.-M. Petit, published in the *Atlas* of the *Voyage de découvertes aux terres australes* (1824), plate 25. Source: National Library of Australia, an7569780.

Figure 5.5: 'Jeune femme sauvage de la tribue Bou-rou-bé-ron-gal, avec son enfant sur les épaules: Nouvelle-Galles du Sud', engraving by B. Roger after N.-M. Petit, published in the *Atlas* of the *Voyage de découvertes aux terres australes* (1824), plate 28. Source: National Library of Australia, an7569787.

This said, Petit's apparent determination to represent the Aborigines as unaffected by colonization is notable for; as Phillip Jones observes, it was a resolve particular to the French at this turn-of-the-century moment.[66] Just as Lesueur produced no sketches of Aboriginal people alongside Europeans or nearby colonial buildings or farms, Petit chose subjects who bore scant, if any, sign of European contact. For example, he did not draw either Bennelong, who evidently was in regular contact with the Frenchmen, or Bungaree, who took part in the Flinders expedition, both of whom were depicted by other artists and depicted moreover in full British attire.[67] It is true that the work of colonial artists and other voyager-artists who visited the colony does generally demonstrate a distancing and selective preference similar to that of Lesueur and Petit;[68] still, other artists did not usually separate native from European as consistently and completely as did Petit and Lesueur. In fact, they did not always avoid the harshness of colonial life in general. In 1793, Malaspina's artist, Juan Ravenet, included Aboriginal people in his sketches of Sydney Town,[69] while his companion Fernando Brambillo included convicts in his 'View' of Parramatta. Furthermore, in 1820, Alphonse Pellion and Jacques Arago, artists on the Freycinet expedition, would most often depict their subjects wearing European clothing – at least shirts and jackets, if not trousers.[70]

Unlike those produced by Petit and Lesueur, both these sets of images accompanied official voyage accounts that engaged explicitly with the matter of the Aboriginal peoples' colonization. Although the Malaspina and Freycinet expeditions sailed in different eras, each pursued explicit territorial and commercial interests on behalf of their respective nations:[71] the Spanish expedition had been carrying out a political reconnaissance of Port Jackson, while Freycinet's mission, like the other expeditions of the Bourbon Restoration era, was influenced by contemporary debate in France about penal transportation.[72] They were therefore both prepared to regard the colony with a critical eye and their anthropological gazes were as open to apparent human limitations as they were to fundamental similarities. By contrast, the work of Baudin's artists represented a national investment in a common human capacity for progress and in the ability of Europeans to guide 'other' societies toward civilization.

The voyagers' view of Aborigines in the colonial field was one that sat ill with the democratic principles of Consulate-era society and with the urgent desire amongst French naturalists and officials to confirm the validity of those principles. Moreover, this transitional stage in the study of Man's natural history, with its combination of continuity and change, had provided the voyagers with particularly broad and demanding guidelines for ethnographic fieldwork. Indeed, it is not so surprising after all that although in 1802 Baudin's was the most explicitly anthropological expedition to visit Port Jackson and though it undertook an unusually prolonged encounter in the colony, in comparison to both earlier and later visiting expeditions it produced exceptionally vague and sparse textual records of the local Aboriginal people. The voyagers' observations of colonized Aborigines were clearly symptomatic of a transformational moment in French politics, society and culture as well as in the study of natural history. They were confused by notions of human rights and civilization as well as by tensions between a sense of duty to gather facts and a sense of personal discomfort in the encounter. When it came to other fields of natural history it would be a different matter but, in relation to their 'anthropological' work, the colonial field proved too confronting for these Consulate-era voyagers.

6 SWANS, FROGS AND RUM: NATURAL HISTORY IN AN 'UNNATURAL' SPACE

Despite their despair about the Aboriginal inhabitants, Baudin's naturalists saw Port Jackson as a promising and rich field sure to advance their contributions to the natural sciences. Péron, for one, newly settled in his accommodation in Sydney with his collections in crates stacked around him and impatient to venture into the countryside, predicted that this sojourn would affect a vital turning point in the expedition's scientific mission:

> New Holland, this country so vast and still so new to the naturalist, would seem to provide us with the most plentiful and important collections, and yet they are almost useless ... Fortunately, our current stopover at Port Jackson must give me the opportunity to most closely observe the animals of this continent and I will neglect nothing in taking the fullest advantage to this end.[1]

The young naturalist and his colleagues, overall, were far more at ease collecting and studying the plants, minerals, animals and even ethnographic objects of Port Jackson than observing the Indigenous inhabitants, whose basic humanity had been too close for comfort. These activities were no less complicated by the particular characteristics of the colonial field, but they were less confronting to the Frenchmen and less disappointing in regard to hopes of national and imperial progress.

In fact, this rejection of one part of the colonial environment and the embrace of the rest forms only one element of how the Frenchmen gave meaning to their scientific collecting and observing at Port Jackson. It will be recalled that the commission of the Institut had intended Baudin to avoid 'sojourns ... in areas already known' – sojourns 'of no use to science'.[2] There was no mention of how colonized areas were placed within this perception of usefulness, but it is implied by the fact that the one colony yet established in the *Terres australes* was not included in Baudin's itinerary. In fact, far from suggesting that the expedition explore the colonial environment, the Comte de Fleurieu instructed Baudin to keep in mind that, just given their proximity to Port Jackson, the waters of Bass Strait were likely to have been already charted by the British and therefore only merited a 'general survey'. It was to the 'unknown' south coast that the expe-

ditioners were required to devote their closest attention: 'a land visited for the
first time offers hopes of new objects to become acquainted with and to collect',
pointed out Fleurieu.³ However, why such land was deemed so important to science was not merely because it was unknown to Europeans and thus offered new
objects but also because it was untouched and therefore provided pristine objects
that could be extracted directly from their natural environment. This original context itself was critical, as Bruno Latour explains. Contemporary savants believed
that if it could be preserved, through the voyagers' careful preparation, storage
and documentation of the objects, then those objects would be rendered 'stable,
mobile and combinable' and they would reach the 'centre of accumulation' in a
fine state for scientific analysis.⁴ Latour concentrates in particular on the process of
this 'accumulation cycle', taking for granted a standard, natural, context that was at
risk within it. However, as Marie-Noëlle Bourguet demonstrates,⁵ it is important
also to investigate the actual collecting and observing *sur la terrain* – from where,
and how, was the material gathered? What significance did the voyagers give to its
provenance? These are particularly crucial questions in regard to a field that did
not correspond with the expectations of metropolitan savants and that called for
improvisation from the voyager-naturalists on the ground. As seen in the expeditioners' ethnographic work, plans laid down in the *cabinet* were not always carried
out quite as intended *sur le terrain*; sometimes, they could not be.

At Port Jackson, or, more precisely, in the broader County of Cumberland,⁶
Baudin's men examined natural features and collected objects in an environment
that had been altered not only by Indigenous land management, as their other
fields had been, but which had been changed more drastically again by English
settlers. It was a multilayered and varied field, comprising towns, farmland and
Indigenous spaces. Would the Frenchmen differentiate between the obviously
'colonized' areas and those that remained 'natural' and least 'known'? Moreover,
would they prefer one type of area over the other? How they viewed the environment would depend largely upon how they experienced and related to it. The
sustained nature of this sojourn and the European comforts to which they had
access during these months provided them with conditions different to those
of typical voyager-fieldwork, which, as Georges Cuvier noted, allowed one to
observe 'interesting objects and living things ... in their natural surroundings,
in relationship to their environment, and in the full vigour or life and activity',
but which was characterized by 'broken and fleeting observations' and the lack
of opportunities for rigorous study or reference to 'books which would tell [the
voyager-naturalist] who had seen that same thing before him'. Yet, neither were
they typical of sedentary work. In his study, explained Cuvier, the observer could
survey 'at his leisure ... all [nature's] products spread before him' while remaining
distant from the distractions of nature and therefore able to think more objec-
tively, freely and profoundly.⁷ At Port Jackson, with onshore accommodation for

five months, Baudin's naturalists had the time and space to consider their findings at some depth, they had wider access to books and more comfortable spaces
in which to study them, and they had a broader range of people – people more
familiar with the local environment – with whom they could exchange knowledge. For these reasons, Port Jackson was as much a small 'centre', a place where
knowledge accumulates and the development of further knowledge is facilitated,
as it was a 'field' on the periphery;[8] and as such, it offered both advantages and
disadvantages to the voyager-naturalists. While Baudin's men discovered some
of their specimens 'in the full vigour of life', they obtained many others through
exchanges with British and Aboriginal acquaintances. In fact, by the time that
the expedition's specimens had been loaded aboard the ships in crates and glass
boxes they had typically passed through a number of hands. And, furthermore,
though the naturalists may have found some sense of distance from nature while
in Sydney Town, effectively, they still inhabited the field. They would each need
to feel their own way around this set of circumstances and opportunities but,
whichever path they took, the colonial field would inevitably challenge their
scientific methods and values.

Its most significant aspect was its politics, for what happened within this
space was not merely collecting and observing but French collecting and observing on British colonial territory. Indeed, the scientific activity during these
months was characterized by a complex intersection of practices and priorities
around knowledge – it was an integral part of the contact history of this sojourn.
When specimens and objects changed hands, when observations were discussed
in the bush, on the street, at the governor's table, traditions in natural history
and gestures of intimacy and power were exchanged as well. The essential nature
of the Aboriginal-Anglo-French relationship during this sojourn is in fact encapsulated in this collaboration. Aboriginal people played a vital role in the colony's
market in natural history objects, but their agency was suppressed in the French
records. Scientific fieldwork is thus represented in these documents as a European activity – an activity pursued by men who possessed, according to Cuvier,
'courage', 'energy', 'the most reliable memory' and 'high intelligence'. And, as it
occurred at Port Jackson in 1802 it is represented, particularly by Péron, more
precisely as a product of Anglo-French collaboration. Of course, the relationship between the colonists and their French guests was not as straightforward
as Péron claimed. The governor and the other colonists made efforts to facilitate
the expeditioners' research because they wished to maintain positive relations
with the French and to claim the honour of contributing to the pursuit of
human knowledge. They also had a direct interest in the results of their guests'
work as the development of the colonial project depended in large part upon
scientific knowledge. In this regard, the arrival of the Baudin expedition – carrying a zoologist, an astronomer, a botanist and a gardener, two mineralogists,

two artists and two geographers – was indeed a propitious event in the colony. The British were eager, therefore, not only to facilitate the Frenchmen's research but to share in the knowledge their guests were producing. Finally, and this is a point that is usually overlooked, King's suspicions about the expedition's political objectives prevented him from giving a free reign to the naturalists' curiosity. As noted earlier, he required Baudin's men to obtain his permission before they ventured into the countryside and arranged colonial officials to guide their explorations of the colonial field beyond Sydney Town. How did the voyager-naturalists go about working within these boundaries? Were their observations shaped or coloured by the imperial tensions – perhaps out of deference to the perceived precedence of colonial knowledge? Did they attribute different values to objects made directly and those obtained second-hand? And did they refer to the knowledge of others when they recorded their observations – if so, what value did they give to that knowledge? Overall, in this context of exchange and in such a busy and shifting landscape, 'known' and 'unknown', or what was either 'new' or 'of no use to science', were ambiguous concepts.

Despite the general transition from the eighteenth century's natural philosophy and collection of 'curiosities' to the more systematic, discriminating and analytical natural sciences of the nineteenth century, most French savants maintained fairly capacious interests. They also often remained conscious of political expediencies; indeed, science and politics were closely associated during this era and some of France's most eminent savants held positions in government. For many, notions of scientific truth and value were flexible, depending upon circumstances, and research was very frequently driven by its usefulness to national interests. The complexities of fieldwork at Port Jackson, therefore, were probably not perceived as quite so problematic by Baudin and his men as they now appear to the historian. In his 'Plan of Itinerary for Citizen Baudin', the Comte de Fleurieu, member of the Institut, declared:

> with regard to the products, he will give his attention to the collecting of those which appear capable of being preserved, and he will apply himself principally to the procuring of useful animals and plants, which, unknown in our climate, could be introduced here.[9]

Port Jackson's colonial field would in fact be an eminently appropriate space in which to judge the potential for Australian plants and animals to be introduced to France. And besides, during the eighteen months before they set sail for the South Seas, Baudin and his men were set a rather grand example of multifaceted fieldwork: the Egyptian expedition.

The Egyptian Expedition, 1798–1801

Baudin's mission, organized by the Institut as representative of an evolution in scientific voyaging and carrying, correspondingly, twenty-two scientific staff at its commencement, in some ways followed precedents set by Bonaparte's expedition to Egypt. The members of the Commission des Sciences et des Arts – who numbered around 150 altogether – had been chosen by the Institut. Among them were some of the Republic's most senior and eminent savants while the majority, like most of those voyaging with Baudin and certainly all of those who remained with him at Port Jackson, were under the age of thirty.[10] Between them, they covered the breadth of the natural sciences.

In Egypt, they contended with similar issues as those Baudin and his men would face in Port Jackson, yet on a much larger scale. This field was generally well-trodden ground, more so than Port Jackson, and it was not an unknown field for French naturalists. Moreover, Egypt was not a port of call for these naturalists – it was the entire mission and, consequently, there was a very significant sedentary aspect to their fieldwork. The savants remained in Egypt for almost three and a half years and established the Institut d'Egypte in Cairo – a centre of knowledge on the periphery. Natural scientists were able both to collect specimens from the field and, without travelling far, then to dissect or prepare and to analyse their specimens in the *cabinet*. And, moreover, as they explored the oriental field, like Baudin and his men, they entered into exchanges with merchants, fishermen and other local people in order to augment their collections. The processes of collecting and studying were indeed facilitated by the particular circumstances of this space; however, producing original knowledge proved difficult. Zoologist and member of the Institut d'Egypte, Geoffrey Saint-Hilaire, wrote to Cuvier: 'what I identify here, the Lacépèdes, the Blochs had observed before me, and have published in Paris and Berlin!' Indeed, by comparison, points out Jean-Marc Drouin, the team led by Alexander von Humboldt and Aimé Bonpland in Latin America was, though less ambitious, far more innovative than the immense expedition in Egypt.[11]

Of course, there was a major, and obvious, difference between the Egyptian expedition and either Humboldt's or Baudin's mission: the Egyptian expedition accompanied a military campaign. As Jean Tulard demonstrates, Bonaparte did not actually consider 'like Alexander, carving out an Eastern empire'; rather, this was about internal politics. Bonaparte wished not only to threaten the British but to distance himself for some time from Paris and he gave the mission a scientific character largely in order to confirm his allegiance with the ideologues.[12] The venture was enormously popular with the French public – orientalism had long been fashionable, the history of European-Oriental contacts revered, and they had recently become yet more so following the writings of the Comte de

Volney. Accordingly, both the work of the savants in Egypt and the public image of Bonaparte himself were romanticized and laden with impressions of historical significance as news of the expedition spread throughout France. Still, it remains true that many of the expeditioners were working not only to advance scientific knowledge but to support the army – for example, by making gunpowder and facilitating relations with the Eyptians. The object of the Institut d'Egypte, moreover, was not specifically, like that of the Baudin expedition, 'to perfect the natural sciences and increase the mass of human knowledge', but rather 'progress and the spreading of enlightenment'.[13] The Institut formed 'the learned division of the army', states Edward Said.[14]

By comparison, although Baudin and his men did not approach Port Jackson with either settlement or governance in mind, and their objectives were more focused on the accumulation than the obtrusion of knowledge, they did closely align themselves with the resident settlers and their colonial project. Moreover, Baudin's expedition and, if in a more general sense, this ambitious scientific-military campaign were both representative of the turn toward a more scientific approach to exploration – a turn that was certainly decisive but which was, also, as Bourguet states, 'an outcome [of the Enlightenment] rather than a rupture'.[15] It must therefore have been with a mixture of traditional and progressive ideas about scientific research, and with impressions of the grandeur of the Egyptian expedition in mind that Baudin's naturalists stepped ashore in Sydney in 1802.

François Péron and the 'noble pleasure of making discoveries!'

Péron, for one, had clearly arrived at Port Jackson hopeful that the relatively 'known' terrain would be more fruitful than the 'unknown' had been. It was known by the English, at least, but not to him – perhaps that was enough. No doubt, like the naturalists of the Egyptian campaign, the task he set himself was not only to discover 'new' animals and objects but to describe what he found in more detail than any previous observers had done.

In any case, before Péron could get his hunt underway, he was ordered by Baudin to compile a report on the zoological results of the first campaign.[16] Most of the collection was stored in the accommodation he shared with Bellefin and the specimens must have filled this space as, during those first days in Sydney, he immersed himself in the work of sorting, cataloguing, describing and recalling his earlier experiences on distant beaches. His report is coloured with details about the physical and emotional context of his explorations, as well as of the animals themselves. During the course of this sedentary work, however, Péron was not isolated from his new field: he and Bellefin received regular visitors at their rooms: visitors wishing to see the immense assortment of specimens.[17] One wonders how often these visitors added their own catches to the collection, or lingered beside it to share their own knowledge about Australian animals. Through the writing of the report, Péron's previous fieldwork would thus have merged with his experience of the colonial field before he had even begun actively to explore that terrain.

No time was lost when the opportunity to begin that exploration finally arrived. It appears to have absorbed far more of his time, in fact, than further cataloguing and description of his subsequent collections. He must have labelled them, but the archives yield no report, like that which he wrote concerning the first campaign, specifically addressing the fieldwork carried out during these five months at Port Jackson.[18] Instead, the specimens gathered during this sojourn are mentioned alongside those from the subsequent campaign in notebooks written, seemingly, back in France.[19] Lesueur, who collected a considerable number of quadrupeds, seems to have been similarly averse to recording the results of his zoological work on paper. Although he drew a number of Port Jackson animals, none of these illustrations appear to have been done *in situ*, as had some of his other zoological drawings.[20] Perhaps, then, both Péron and Lesueur put the business of exploring, collecting and observing, as well no doubt as the amusement of social activities, above the sedentary aspect of natural history work.

Péron did range widely across the County of Cumberland. His first excursion was to Botany Bay – the *lieu de mémoire* of the 1788 disappearance of the La Pérouse expedition.[21] This foray was no doubt as much a pilgrimage as a natural history excursion. Péron disapproved of the 'arid and sandy' landscape of Botany Bay, which, he opined, 'does not appear suitable for any type of cultivation' and where indeed there were no European settlements to be found.[22] Furthermore, while in his account he names some of the trees seen there and mentions the 'unfortunate tribes who live on these miserable shores,'[23] he makes no mention of objects collected. His notes reveal that Mrs Paterson – the wife of William Paterson, lieutenant-colonel of the New South Wales Corps – gave him some shells from Botany Bay,[24] but there is no evidence that he collected anything there himself. Given the tone of his account, Péron may have been attempting to fulfil the role of the 'manly' and courageous naturalist in the field;[25] however, if that is so, he fell short by providing no material discoveries as evidence of his accomplishment. Interestingly, both his attitude concerning the agricultural environs of Parramatta and Castle Hill as well as the results of his fieldwork there, were very different.

Péron was enraptured with this green and fertile landscape and, in particular, with all that the English had done with it. He waxes lyrical about the settlers' agricultural accomplishments, the 'pretty dwellings' and the township of Parramatta, and refers only fleetingly to the forests that separated the settlements or the rivers that ran through them. As his narrative skips from merino sheep to his zoological collection, one gets the impression that the specimens had been simply plucked out of this colonial Eden – the fruits, if not of the naturalist's

courageous confrontation with the natural world, then of its European civilization. Péron emphasizes the size of the collection, and that part which came from
the Parramatta region alone was considerable. He records finding 150 new species of insects, including forty butterflies – which for one excursion constitute an
impressive proportion of the expedition's final total of 880 new insect species.[26]
The lizards he found made up a quarter of the total number of new lizard species
discovered by the French in Australia. And, notably, this was the only location
in Australia, including Tasmania where Péron found frogs, tree frogs and – apart
from one 'very small' specimen found under a log in Geographe Bay – toads.[27]
Similarly, the platypus, emu and long-necked turtle were to be the only specimens of their kind in the expedition's collection. Péron was proud that these last
three animals were 'sent by me to the Muséum'.[28]

Péron attributes all of these finds to himself, but, of course, he had not been
alone in the field. During his first visit to Parramatta, he was accompanied by
his companion Bellefin. Although the naval surgeon's journal is not available, it
may be assumed that he collected alongside Péron and that, similarly, his medical skills were applied to the tasks of preparing, preserving and evaluating the
specimens. He and Péron had been assigned a guide – a sergeant of the New
South Wales regiment – whom Paterson had ordered 'to procure us the means of
pursuing our studies as fully as possible'.[29] Part of his duty may have been actually
to contain the naturalists' explorations, for there was terrain which the English
were determined to 'discover' themselves. For example, King flatly refused the
requests of Péron, as well as of mineralogists Bailly and Depuch, to venture into
the Blue Mountains – even in the company of colonial authorities.[30] Nevertheless, the sergeant's role must otherwise have facilitated the naturalists' research.
He obtained accommodation for them with some of the wealthiest proprie-
tors in the colony, who were not only able to provide comfortable lodging and
to facilitate multiple, unhurried excursions but also, no doubt, to share useful
knowledge of the land and its productions. Sojourns with settlers such as James
Larra, at whose inn Péron and Bellefin 'were consistently served with an elegance
– a richness, even – that we should never have believed possible on these shores',[31]
could only have reinvigorated the naturalists and fuelled their research efforts.

In fact, even though Péron was usually vague about its results, collaboration
was a central theme in his representation of this colonial field. His narrative of the
Port Jackson sojourn is sprinkled with references to the support offered by the
governor, the assistance provided by prominent landowners, the interest in his
work shown by the inhabitants of Sydney Town and the participation of English
collectors in his fieldwork. He was particularly enthusiastic about his experience
of sharing the field with Patterson – gentleman naturalist and correspondent of
Banks, as well as lieutenant-colonel – and George Caley, Banks's official natural
history collector in the colony. He wrote in the *Voyage de découvertes*:

How interesting it was to spend several days traversing these areas, so rich in objects new to Europe! With what enthusiasm did we vie with one another in the noble pleasure of making discoveries! And with what affectionate generosity did my honourable collaborators bestow upon me all the objects that my own searches had failed to uncover![32]

Once more, Péron is more concerned with the context in which his collecting took place than in exactly what he collected. Of course, he kept inventories and notes of specimens obtained but, conversely, they lack context. The gentlemanly, civilized nature of the activity, and perhaps the respectable company in which it took place, were probably intended to give the unnamed specimens greater value. The claim about their scientific significance is unsubstantiated; instead, they are put forward as representations of the 'honour' and 'nobility' of European 'discovery'. No mention is made here, or indeed anywhere else in the official account, of Aboriginal involvement in Péron's fieldwork. In an unpublished document, Péron describes an Aboriginal man named Ourou-Mare, called 'Bulldog' by the settlers, whom he 'kept' with him throughout the sojourn and whose skill at catching lizards and snakes he keenly praises.[33] Aboriginal people in fact played a central role in what was a thriving natural history economy at Port Jackson, and Caley himself relied heavily in his fieldwork on the knowledge and skills of his own Aboriginal companion;[34] it is likely, then, that Ourou-Mare caught those reptiles for Péron's zoological collection. However, Péron clearly crafted one narrative for New Holland's 'savages', as shown in the previous chapter, and another for 'civilized' Europeans. Perhaps seeking to compensate in the context of natural history for the fact that this colony was no 'unknown' beach, quite possibly remembering the prestige of the Egyptian expedition, Péron chose to promote the colonial field as a triumphant European space.

At Port Jackson, as Margaret Sankey demonstrates, Péron's scientific gaze broadened, to encompass the development and administration of the colony itself.[35] That is not to suggest that his zoological work was put aside; in fact, it is clear that it was continued with enthusiasm. However, Péron sought to highlight his familiarity with the British colony – in fact, as Sankey suggests, to possess it with his knowledge – and to give greater significance to his natural history work at Port Jackson by presenting it in that context.[36] His collecting was strongly affected by the culture of the colonial project and, in particular, by the Enlightenment concept of improvement through scientific knowledge.[37] In actual fact, Péron turned that notion around in his narrative – by representing knowledge as a product of improvement.

Théodore Leschenault: Botanizing the Colony

This was certainly not an approach taken by his fellow scientific voyager, botanist Leschenault. The one available record of the Port Jackson sojourn, authored by Leschenault himself – a report on the vegetation of New Holland and Van Diemen's Land – shows that the young French botanist believed the colonists' attempts to cultivate the land were not bringing any improvement to the environment at all but, in fact, were only proving unproductive for farmers and detrimental to the land. He explained:

> When accidental causes have not *enriched* the soil, the farmer is often disappointed in the hopes that had been raised in him by a country covered in fine forests – the slow, gradual product of several centuries of growth, utterly undisturbed by human industry. Few years are needed to exhaust a land that he has painfully cleared. In the area surrounding Parramatta, I came upon a number of these farms that have been abandoned. After cultivation, the soil (after its return to Nature), is no longer covered by anything but puny bushes and a species of *Saccharum*, a dry, course, graminaceous plan, not suitable for feeding live-stock.[38]

This concern for the destruction of native forests at the hands of European farmers would have been of interest to Baudin's friend at the Muséum d'Histoire Naturelle, botanist and conservationist André Thouin. Moreover, it was a concern stemming more from a scientific than an imperialist point of view.[39] Leschenault looked at how the settlers were using and changing the land and its productions not in order to admire the march of civilization but to further his understanding of the Australian environment: that is, to learn about the properties of the plants and the possibilities and limitations inherent in the land. Although his insight was limited by the common assumption that the land had been untouched, unmanaged prior to European settlement, the colonial nature of the field considerably facilitated his investigations to this end.

Leschenault did emphasize in his report the view that this scientific terrain was 'already known'. The plants of this region, he declared, had been almost all described by English botanists. Accordingly, he went on, as he explained himself, to specify, describe and categorize only those particular plants in his collection which he deemed new and, therefore, of interest: a species of *Dianella*, another of *Exocarpos* and several plants of the legume and myrtle families.[40] This more disciplined and scientific approach further sets Leschenault's work at Port Jackson apart from that of Péron.

However, if this field was not untouched, that perceived disadvantage was offset by the various benefits – such as accommodation, transport and prolonged time ashore – which the sojourn offered. Each of the discoveries Leschanault listed had been collected during an excursion to the foot of the Blue Mountains – a distance inland that he could not have reached during any of the expedition's other stopovers on the Australian coastline. Furthermore, while specimens new to European science may have been difficult to find, no doubt most of the plants and seeds were, like Péron's zoological specimens, new additions to the cabinets of the Muséum. Indeed, the Port Jackson sojourn produced an immense botanical collection. Towards the end of the sojourn, embarked aboard the *Naturaliste* for her homeward voyage were: 79 tubs containing a total of 800 individual living plants (of around 250 species), 3,560 dried plants (of around 900 species) and, as mentioned above, 3 crates of seeds. It is not possible, based on sources currently available, to determine precisely what proportion of this collection had been gathered at Port Jackson and how much of it was the result of the first campaign. However, Michel Jangoux makes an interesting observation: as the work he had been undertaking at the Muséum d'Histoire Naturelle in Paris suggests, most of the specimens collected by the Baudin expedition in Australia were from the County of Cumberland.[41]

Like Péron's collection, Leschenault's crates of plants and seeds were gathered alongside fellow naturalists, under the supervision of English guides, and no doubt – in some cases – by hands other than his own. British records show that, during the *Naturaliste*'s first visit to Port Jackson, Leschenault carried out fieldwork alongside the eminent botanist of the Flinders expedition, Robert Brown. Presumably, they worked together again during the young Frenchman's second sojourn at Port Jackson – Brown was still in Sydney and it was to be almost another month before he departed aboard the *Investigator*.[42] Interestingly, however, Leschenault himself makes no mention of this connection or, indeed, of collaboration of any sort. His report is focused almost entirely on the practical uses and limitations as well as the scientific significance of Australia's vegetation, diverging only briefly to discuss his view of the Aboriginal people and their relationship with the land. It suggests a systematic approach on the part of the voyager-botanist and, moreover, an attempt to represent Port Jackson first and foremost as a botanical field rather than as a European colony.

Leschenault no doubt believed that representing his work in this way, without reference to the assistance and exchanges that it had involved, lent greater scientific authority both to his specimens and observations as well as to himself as a voyager-botanist. Of course, in its style and content, Leschenault's report may be seen further as asserting superior French knowledge over the bumbling efforts of English farmers and less than thorough fieldwork of colonial collectors.

Charles Bailly and Louis Depuch: Useful Mineralogy

To ensure that his efforts at Port Jackson would be deemed worthwhile, Leschenault needed to advance a body of botanical work that the English had been developing since 1788. In the discipline of mineralogy, by contrast, his colleagues faced no such challenge. Apart from the Abbé Mongez of the Lapérouse expedition, who in February 1788 advised Governor Phillip on the suitability of local white clay to 'make good China',[43] Bailly and Depuch were the first mineralogists to set foot in Australia. This meant not only that they would be certain to make discoveries, and therefore to contribute significantly to the success of the expedition and to French scientific knowledge, but also that their colonial hosts at Port Jackson would deem their expertise of particular value. The relationships that Bailly and Depuch would enjoy with the colonial officials would therefore have been somewhat different from that experienced by Leschenault and Péron and, certainly, the way in which the young scientists would represent their work in this field would also differ distinctly.

The mineralogical records currently available are both authored by Bailly – Depuch died during the return voyage and his papers are yet to be recovered. Nonetheless, Bailly's reports are written with reference to collective research, undertaken by the author in collaboration with the more senior Depuch.[44] They comprise an account of the excursion Bailly and Depuch made through the environs of Toongabbie and Hawkesbury, to the edge of the Blue Mountains, and a descriptive inventory of the specimens they collected and were given during the course of the Port Jackson sojourn.[45]

The first of these documents demonstrates that Bailly and Depuch were not solely interested in the region's geological make-up. Using a narrative style similar to Péron's, though more coherent, Bailly recounts the events of the mineralogists' excursion and lays out the observations that they made along the way – observations that did mainly relate to the land but which otherwise focused squarely on the progress of the colonial project. He devotes particular attention to the town of Toongabbie and, in particular, to the signs of order and wealth it revealed and the agricultural prosperity it represented. Thus, Bailly gives much more attention in this report to extra-disciplinary matters than does Leschenault in his own – Leschenault only writes about the colony through the lens of botanical analysis – yet he neither romanticizes the colonial project nor glosses over his scientific work as Péron tends to do in the *Voyage de découvertes* narrative. While the mineralogists evidently saw merit in overtly combining the political with the scientific, it does not appear that they did so in order to boost the import of their mineralogical research but, rather, that they saw the colonial context both as integral to the nature of their findings and as worthy of comment in and of itself.

In fact, it is apparent that the mineralogists' knowledge was of greater value to the colonial officials than it was to the scientists at the Muséum. Bailly makes direct mention of the assistance that was given to him by the colonial administration: Governor King sent letters of recommendation and a request to host the Frenchmen to Mr Arndell – surgeon and first magistrate at Hawkesbury, Arndell gave Bailly and Depuch accommodation, a guide and local advice. Either King or Arndell also provided the mineralogists with an interpreter, equipment and supplies.[46] Similar assistance had of course been offered to the other French naturalists as well but, in this case, the contribution of the English consisted almost entirely of support and facilitation rather than, also, researching side-by-side with the French as fellow naturalists. True, Bailly's inventory shows that King and Paterson contributed several mineral samples to the expedition's collection;[47] however, the colonists were generally unqualified to share the task of mineralogical analysis with Bailly and Depuch. Instead, they sought to benefit from the Frenchmen's knowledge. Although King would not permit Depuch and Bailly to extend their research into the mountain range, he did give Depuch the opportunity to study some rock samples brought back from Francis Barrallier's expedition into the Blue Mountains. Following his analysis, Depuch was able to inform King that the expedition had not reached the centre of the range, as the rocks they had collected did not include samples of its granitic core.[48] Based on analysis of mineral samples from the banks of the Nepean River, as Mayer explains, Depuch had earlier been able to conclude that the core of the Blue Mountains was of an ancient composition.[49] Furthermore, their research in the colonial field also revealed the existence of Triassic deposits of sandstone and shale, as well as indications of valuable coal layers – finds that are bound to have been of interest to participants in the colonial project.[50] In fact, it is worth noting that, of all of the research undertaken by Baudin's naturalists at Port Jackson, it was only that of Bailly which King referred to specifically in his correspondence with Joseph Banks. He wrote:

> While the *Naturaliste* was here the mineralogist made experiments on the ferruginous stones that abound here. He says they contain too small a portion of iron for working, but that a profitable substance might be got from them for glazing porcelain.[51]

This said, the discoveries made during the voyage by Bailly and Depuch, and, it should be noted, also by Péron,[52] made little impression on the members of the Institut National. It seems Jussieu even overlooked the work carried out at Port Jackson when he reported in 1804:

> It is not surprising that, from research carried out mainly on desert or heavily wooded coasts, which offer neither rising mountains, nor ravines, for learning about the composition of the land, nor any work of exploitation, the mineralogists Depuch and Bailly could collect only a small number of mineral samples, insufficient to give an exact idea of the geology of this country.[53]

It is possible that the mineralogists' research at Port Jackson had been influenced more by the concerns of the English colonists, and indeed by Bailly's and Depuch's interest in the colony, than by the demands of science.

Charting the Settlement and the Stars

It was certainly most striking in relation to the mineralogical research but, colonial officials had been eager to participate in the French voyagers' production of knowledge throughout the duration of the sojourn. When it came to the expedition's astronomical work, King literally demanded a share in the results. He required Bernier to send to him each day, via one of the soldiers of the New South Wales Corps, a copy of all the data he had collected. This requirement may not have significantly influenced the quality or nature of Bernier's work, if at all; however, it is worth noting for how it highlights the shared and multi-purpose nature of the research carried out by voyager-scientists in the colonial field. While King may have demanded regular reports from Bernier due to a particular interest in astronomical research – for instance, he may have used the data to improve the accuracy with which the longitude of Sydney was known – it is worth considering, given that there is no evidence that such requirements were made of the other French scientists, that King's rationale was partly territorial. That is, that the colonial administration had a right to any data collected on colonial territory. And, as Bernier spent the majority of his time during the sojourn in the observatory tents, situated on Bennelong Point, King did not have ready access to his findings as he did to those of the other French fieldworkers. In lieu of casual exchanges of knowledge, regular reports provided a means of gaining access to the knowledge that was obtained within this colonial space.

Bernier, in his turn, made no claims upon the British colony. His journal entries diverge rarely from astronomical observations. By contrast, however, the French geographical work was focused distinctly, and quite naturally, on attaining a thorough familiarity with the layout and construction of the colony at Port Jackson. Although British maps had been available in France since the early of years of the colony, the Baudin expedition's visit to Port Jackson provided the first opportunity for Frenchmen to produce, from personal observations, their own plans and charts of the region. As acknowledged in their titles, the two main charts were based on existing British charts: 'Plan du Port Jackson', which was based on those of former governor John Hunter[54] and the 'Plan du Comté de Cumberland', which was drawn up from 'les Cartes Anglaises' (see Figure 6.1 on p. 115).[55] Thus, from the beginning, these documents were to constitute a combination of British and French knowledge. However, the final result was not just a layering of French over British representations, but of French artistic, cartographic and geographical views over an official colonial view: the expedition's charts were based on observations made at Port Jackson by Boullanger, Lesueur and Louis Freycinet.

If the French version of the map of Port Jackson is compared to those drawn up for Hunter,[56] it is evident that the British charts were used only to provide a

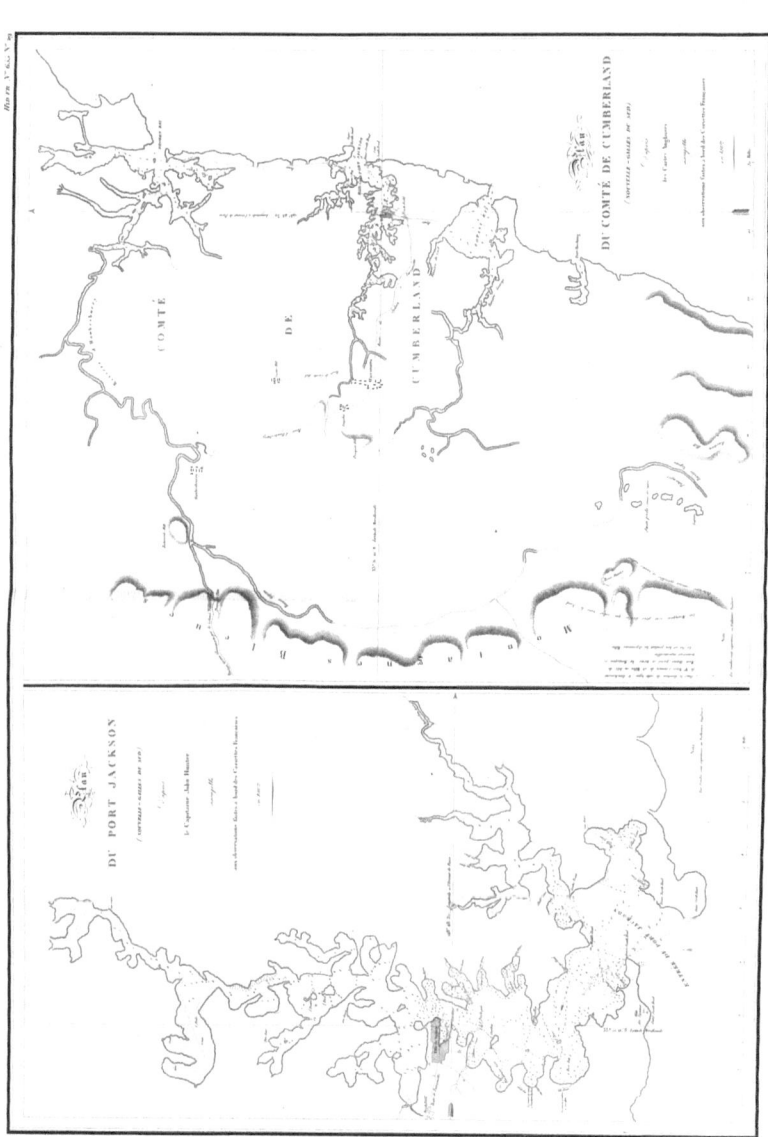

Figure 6.1: 'Plan du Port Jackson (Nouvelles Galles du Sud) d'après le Capitaine John Hunter assujetti aux observations faites à bord des Corvettes françaises en 1802'; L. Freycinet and 'Plan du Comté de Cumberland (Nouvelles Galles du Sud) d'après les Cartes Anglaises assujetti aux observations faites à bord des Corvette Françaises en 1802; in the *Navigation et géographie, Atlas of the Voyage de découvertes aux terres australes* (1812), plate 29.

Source: National Library of Australia, MAP RaA 2.

broad outline. Based on observations made aboard the French ships in 1802, as the titles indicate, the French maps provide detailed contemporary representations of the colony. The map of Port Jackson offers a closer view than the maps upon which it seems to have been based and therefore includes additional details such as the names of numerous points and coves along the coastline, locations of fresh water sources and Sydney Harbour's point of longitude from Paris. The map of the County of Cumberland shows the roads linking each of the colony's townships, indicates the size and shape of each town, as well as features such as waterfalls and rivers at the edge of the mountains and some notes about the fertility of land at different points near the mountain range. The Frenchmen's interest in the colony, however, was not limited to the layout of the country and the quality of its soil. Their map of Sydney is immensely detailed.[57] It was sketched by Lesueur according to bearings taken by Boullanger, and charts in fine detail the area from the southern edge of Palmer's Cove to the northern edge of the township, and from Sydney Cove inland to the village of Brickfield. Civil and military structures, docks and dockyards, gardens, even the cemetery, are carefully depicted and cross-referenced.

While, as shown, there is a great degree of detail in the maps and illustrations of the colony at Port Jackson, there is as yet no evidence that the geographers, Boullanger and Faure, undertook any other geographical studies of this region during their five months ashore.[58] In fact, Faure does not appear to have recorded any geographical observations during the course of this sojourn at all – a circumstance which merits investigation. The importance of charting ports, particularly those that could hold military importance, had been made clear to Baudin, and presumably to his staff as well.[59] Although a reconnaissance of Port Jackson itself had not been planned, access to detailed plans of Britain's flourishing colony would obviously have been of great strategic interest to French authorities. It would indeed not be surprising if, as Jean-Paul Faivre remarks, the French plan of Sydney was that which 'Napoléon ... kept upon his desk'.[60] The opportunity to chart the British colony apparently engrossed Boullanger to the exclusion of the sort of comprehensive geographical studies that he normally carried out in onshore fields.

Objects over Humans: Ethnographic Collecting

Nonetheless, it was the expedition's anthropological study that was affected most deeply by the colonial environment. As noted earlier, disappointment about the Aborigines of Port Jackson did not prevent the Frenchmen from significantly augmenting their collection of ethnographic objects and records of specific aspects of Aboriginal culture. Lesueur and Bernier together produced the first European notation of Aboriginal music. One of the pieces of music – which the

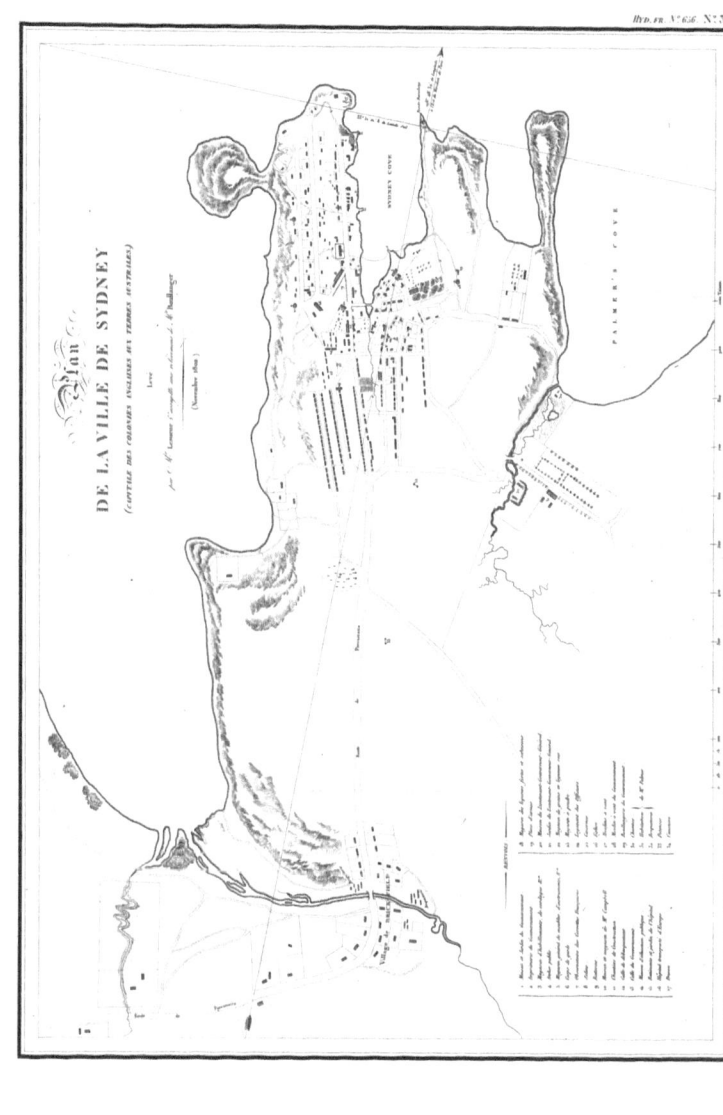

Figure 6.2: 'Plan de la ville de Sydney (capitale des colonies anglaises aux Terres australes) levé par M. Lesueur et assujetti aux relèvements de M. Boulanger. Novembre 1802; by C.-A. Lesueur, in the *Navigation et géographie, Atlas* of the *Voyage de découvertes aux terres australes* (1812), plate 30. Source: National Library of Australia, MAP RaA 2.

French entitled 'Cri de ralliement'– is in fact the well-known 'Coo-ee', which of course would eventually become part of the popular Australian identity.[61] And Lesueur also made copies of Aboriginal rock art and produced a number of fine drawings of Aboriginal tools and weapons (see Figures 6.3, 6.4 and 6.5 on pp. 119–21).[62] In fact, eighteen years later, English explorer Phillip Parker King referred admiringly to Lesueur's sketch studies as the principal reference on the subject of Aboriginal weapons. Following an encounter with a group of Aborigines at Port Bowen (now Port Clinton, Queensland), King described the boomerang that one man held and noted:

> Boomerang is the Port Jackson term for this weapon, and may be retained for want of a more descriptive name. There is a drawing of it by M. Lesueur in Plate 22 Figure 6 of Peron's Atlas; it is there described by the name of sabre a ricochet. This plate may, by the way, be referred to for drawings of the greater number of the weapons used by the Port Jackson natives, all of which, excepting the identical boomerang, are very well delineated.[63]

However, it is notable that most of the items added to the ethnographical collection at Port Jackson were items received as donations, not gathered by the Frenchmen themselves. Baudin was particularly happy to receive a donation that would, in the end, constitute over three-quarters of the entire final collection: 160 Oceanic objects, bestowed by English explorer and merchant George Bass and intended for the museum of the Société des observateurs de l'Homme.[64] The vocabulary and grammar of the 'savages' of Port Jackson were also donations – one, again, from Bass, and the other from Paterson.[65] Finally, an inventory compiled by Péron shows that more ethnographical objects were added to the collection during this sojourn than at any other stage in the voyage; however, it does not indicate if any of the objects were donations or obtained directly via exchanges with Aboriginal people.[66]

The French voyagers made little attempt to give meaning to their ethnographical collection. Péron evidently deemed geographical provenance of objects as critical, for he made note of the region in which each piece had been collected, but not necessarily their source. In fact, it also worth noting that none of the expeditioners gave descriptions of the uses of these items or of their importance in Indigenous life in their journals or other papers; Péron's listed them, and they were added to the collection. Of course, there was little explanation either Péron or any of his colleagues could have given to the Pacific Island objects donated by Bass. But they had had ample opportunity at Port Jackson to record observations about Aboriginal music, art, weapons and tools.

Figure 6.3: 'Nouvelle Hollande: Nouvelle-Galles du Sud. Musique des naturels', notation made by C.-A. Lesueur and A. Bernier, in the *Atlas* of the *Voyage de découvertes
aux terres australes* (1824), plate 32. Source: National Library of Australia, FRM F979
(Atlas), p. 32.

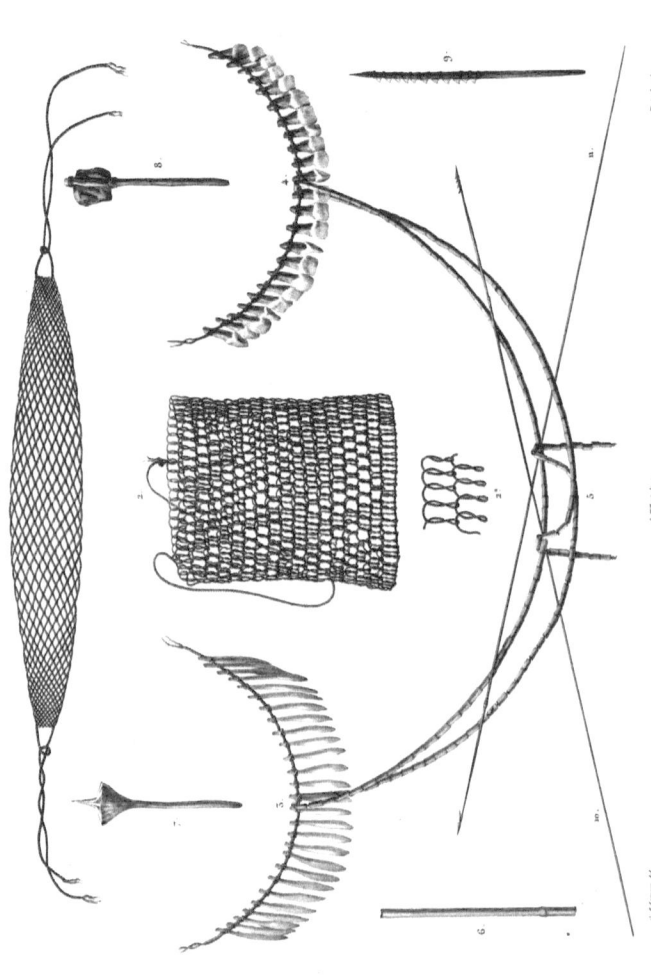

NOUVELLE – HOLLANDE : Nouv.ᵉˡˡᵉ Galles du Sud.

ARMES, USTENSILS ET ORNEMENS. (D'après l'explication des Planches.)

De l'Imprimerie de Langlois

Figure 6.4: 'Nouvelle-Hollande, Nouvelle Galles du Sud, armes, ustencils et ornemens', engraving by F. M. Testard after C.-A. Lesueur, in the *Atlas of the Voyage de découvertes aux terres australes* (1824), plate 29. Source: National Library of Australia, an7569788.

NOUVELLE - HOLLANDE : Nouv^{le} Galles du Sud.

DESSINS EXÉCUTÉS PAR LES NATURELS.

1. Espèce de Morène . 2. 4. 5. Diverses figures du Dieu des Montagnes bleues . 5. Squale barbu . 6. Kangurou . 7. 8. 9. Poissons .

9. Espérance de Langlois.

Figure 6.5: C.-A. Lesueur, 'Nouvelle-Hollande, Nouvelle Galles du Sud, dessins exécutés par les naturels', engraving by F. M. Testard after C.-A.Lesueur, in the *Atlas* of the *Voyage de découvertes aux terres australes* (1824), plate33. Source: National Library of Australia, an7569799.

Perhaps they were simply trying to be good fieldworkers – gathering material to be passed on for analysis by metropolitan naturalists. The members of the Société des Observateurs had provided Baudin with a guide to the types of objects they wished him to gather – objects that would 'assist in the formation of the special museum of the Société' – and it did not in fact suggest that the collection should be accompanied by descriptions or explanations. Indeed, it was assumed that the objects would speak for themselves in relation to universal meanings about degrees of civilization and human nature. Documents illustrating 'the various states' of music and drawing, for example, were believed to 'open up fresh sources of pleasure, happiness or instruction'.[67] All the same, it is interesting to note that not only did the expeditioners not attach explanations to specific items in their collection, but they also made markedly fewer references to material, spiritual or artistic culture in their general accounts of Port Jackson's Aboriginal people than they did in their reports of contact either prior to or following this sojourn.

Péron had written at length about Aboriginal burial sites, several men recorded detailed observations of huts and 'monuments', and Ransonnet would describe, with colour, his acquisition of some weapons and tools.[68] In fact, put alongside the reticent observations recorded during these months, the Aboriginal items from Port Jackson, from the 'Coo-ee' to the spears listed in Péron's inventory, stand as peculiarly detached from their human context – a further manifestation of the Frenchmen's tendency to keep the Aborigines of the colony at a distance.

Yet, Baudin dearly prized the entire ethnographic collection. Instead of sending it, with the other natural history specimens on the *Naturaliste* directly to France – which would have had it delivered to the Société within around six months – Baudin kept it with him aboard the *Géographe*, for the duration of the second campaign. Presumably, he wished to deliver it, in person, to the Societé in Paris.

The Scientific Captain and the Dispatch of the Natural History Collection

As for the vast remainder of the expedition's natural history collection, gathered during the course of the first campaign and these months in port, it was Baudin's wish that it be sent to the Muséum without further delay. In fact, he had contributed much to this collection himself. Some of the expedition's most prized acquisitions – including emus, a spotted quoll and a pair of black swans – were the result of the captain's dealings with English settlers during this sojourn. Many had been traded for quarts of rum; for example, in the month of September alone he exchanged three quarts of rum for a 'native cat', three and a half pints for a fish and a gallon for a swan.[69] Many others, like the ethnographic objects from Bass, were generous donations. Mrs Paterson, the lieutenant-governor's wife, donated a collection of 400 shells of various species and taken from regions in New Holland, the Pacific and America.[70] And some Baudin had obtained during the course of an excursion in the company of Governor King, which, he proudly boasted to Jussieu, took him 'beyond the furthest areas known to the English'.[71] If this comment seems to indicate a sense of intellectual rivalry on the part of the French captain, that impression is confirmed by another of his remarks to Jussieu concerning the pines from New Zealand and Norfolk Island that he had acquired: they would no doubt be 'most appreciated because no other European nation has been been able to procure them', he pointed out.[72] Perhaps this sense of rivalry further fuelled Baudin's determination to deliver the natural history collection to his superiors as quickly and safely as possible. In any case, the Port Jackson stay gave him the opportunity to attempt this with as much care as possible. He replaced the *Naturaliste* with a new vessel, purchased with Governor King's permission because it would contribute to the pursuit of science and was intended as a geographical tool for the second campaign, and ordered his men to load all of the botanical, zoological and mineralogical specimens aboard the original consort ship. Péron and his colleagues spent two weeks arranging the placement of live animals and plants and of crates and glass boxes aboard the *Naturaliste* – they needed to reach their destination in sound condition and in order. Organization alone, however, would not suffice in Baudin's view. Before Hamelin set sail aboard the *Naturaliste* with his precious cargo, Baudin furnished him with a set of lengthy and detailed instructions on how to care for the specimens.[73]

As it turned out, Hamelin was not the conscientious scientific captain that was Baudin and a large part of the collection was ruined on the homeward voyage;[74] however, while he was delivering specimens to the scientists in Paris, Baudin and his naturalists would be busy gathering an even larger collection on the least familiar points of the Australian coastline.[75] This sojourn in an Australian colonial field, as Sankey reveals, had enabled the captain and his naturalists to re-evaluate their scientific objectives.[76]

Indeed, two seasons on shore, with the facilities of a nascent colony at hand, had certainly provided Baudin's scientific staff with the space to consider and approach its fieldwork from a different angle than that permitted during beach excursions and stretches at sea. On the terrain of the colonial field, these turn-of-the-century voyager-naturalists had used a variety of methods to further their research. On the ground, augmenting their collections was of the utmost importance to them all – in town, on farms and in forests they bartered, collaborated, shared and discovered specimens both already known and yet unknown to natural historians. Their styles in the field are not easily categorized, yet one can observe in their work at Port Jackson the shift towards a clearer compartmentalization of disciplines; that is, with the exception of Baudin and Péron, who contributed to a range of areas. Moreover, while their manner of collecting may have been fairly indiscriminate, there is no doubt that they were all concerned about the contributions they were making to French knowledge.

The precise meaning each gave to his scientific findings depended heavily upon his own view of the colony – consider, for instance, the approach of Péron, in zoology and other fields, by contrast with that of botanist Leschenault. The findings were also given value, most notably in the discipline of mineralogy, by the colonial authorities on whose territory they were made. In fact, one of the most important characteristics of the French naturalists' work at Port Jackson is that it was carried out in a public space. Granted, the research of voyager-naturalists was very often carried out under the watchful, usually distant, eye of coastal Aborigines; but, at Port Jackson, its processes, results and representations were affected most particularly by the surveillance of English colonists. They had a strong interest in their visitors' research – it corresponded with their zeal for trading curiosities and, more importantly, it had the potential to benefit the colonial project. It was a valuable commodity and one gained on English territory. Thus, in part because of the influence of their hosts but also because they were deeply interested in the British development of this environment, the Frenchmen themselves related their work most often to issues of colonial endeavour. In the end, they found a special significance for their findings from this unscheduled sojourn, in this complex space. As the research in Egypt had been imbued with the romance of an encounter with the Orient, of progress and of French civilization, Baudin and his naturalists would present to the Institut not just antipodean specimens, objects and data, but the natural science of an Australian colony.

7 BAUDIN'S 'NEW EXPEDITION'

'I beg you not to forget me', wrote Baudin to Jussieu, 'and I am going to do all I can to gather a new collection just as large as that you will receive by the *Naturaliste*'.[1] Indeed, he had spent the entire sojourn planning and preparing to achieve just this, along with more thorough and accurate charts and more comprehensive observations. John Dunmore notes that these months on shore had allowed Baudin the time and space to evaluate the first part of the voyage,[2] and of course, they did. But his behaviour on the south coast earlier that year shows that he had already been dissatisfied with his work and ready to renew his efforts when he made the decision to visit the British colony. In port, Baudin not only appraised the preceding voyage but constructed a new expedition. In the words of Horner, the commander discarded 'the superfluous material' and used all he had learned in Australian waters until that point to build a 'leaner, more efficient instrument for carrying out his instructions'.[3] Horner gives a neat summary, however, the process of reassessing the objectives of the voyage, the mechanics of disassembling and reconstructing the expedition, indeed, the resolution to begin afresh from Port Jackson at all, were more significant than these comments suggest.

It must be recognized from the outset that a *new* voyage was not necessary – it was a choice. Although Baudin himself was unsatisfied with the results, the expedition had in fact fulfilled the basic expectations of the government. The only part of the itinerary that had not yet been attempted was Australia's north coast – had he wished, Baudin could have followed his English counterpart, Flinders, and sailed directly there, thus completing his exploration. But he chose otherwise. 'Upon departing from Port Jackson', Milius later told the Minister of Marine, 'the commander was so earnest that he was determined ... to sacrifice all the time necessary, etc., even his life, in order to completely fulfil the object of his mission'.[4] How, in precise terms, he had prepared to achieve this, cuts to the heart of the Australian voyage, both in regard to Baudin's personal drive as a scientific captain and to the value of this venture to French intellectual and political authorities. The expedition, as a whole, was about gaining knowledge. Its re-construction, then, is clearly just as meaningful as the collecting and observing carried out by its scientific staff on the ground.

The most vital element of the new voyage was, of course, its itinerary. Like the 'Plan of Itinerary',[5] it comprised the south, west and north coasts of Australia. Baudin planned to rechart certain areas that he felt had been imperfectly done, to add topographical studies to the expedition's existing charts and to allow his naturalists to venture onto beaches that he had previously been unable to approach due to inclement weather. The itinerary was sketched out in letters to Jussieu and the Minister of Marine shortly before leaving Port Jackson.[6] Sailing directly to Bass Strait, Baudin would first survey King Island, newly discovered by British fishermen, and then Kangaroo Island, the south coast of which neither Matthew Flinders – who discovered it only shortly before encountering Baudin – nor the French had examined. From there, the *Géographe* and the *Casuarina* would return to the St Peter and St Francis Islands so that Baudin could assure himself of the 'direction of the continent' from that location. The expedition would next perfect its survey of Geographe Bay, the expedition's first 'discovery' upon reaching Australia in 1801, before searching for the 'isles du Romarin', first charted by William Dampier. These islands had not been included in the original itinerary. Baudin explained to Jussieu that, because of the inaccuracy of the maps provided to him in France, he had been unable to find the islands during the first campaign and therefore it was 'for the perfection of geography' that he wished to determine and chart their exact position.[7] From there, his ships would sail north to re-chart the north-western coastline, as Baudin felt that the expedition's first surveys had 'lack[ed] the perfection necessary for safe navigation'.[8] The final stage of the campaign, in accordance with Fleurieu's plan, was to include the north of Australia as far east as the Gulf of Carpentaria.

Many of these were places to which Fleurieu had instructed Baudin to pay particular attention; however, in the case of Bass Strait for instance, others were areas to which Baudin gave a significance that the eminent geographer and past Minister of Marine had not. During the course of the first campaign and then in further detail during his stay at Port Jackson, Baudin had learned of islands in Bass Strait about which the French authorities had been ignorant and that he himself had not yet seen. These he added to his itinerary. He had learned too, of course, about British interests in that region. Although, according to Fleurieu it was not worth carefully charting an area already made 'known' by British maps, Baudin had come to the conclusion that it would be judicious for the French to have their own, more accurate, record. His reasoning to this end, though, was clearly complex, for there is a curious omission in the south-east section of his itinerary: Port Philip Bay. This was a very recent British discovery and it would have considerable strategic significance for whichever European nation eventually settled it. Governor King had already appealed to the Admiralty to form a colony there and no doubt he made his intentions known to Baudin. Yet, although the French government would almost certainly have prized French-made charts of the bay – John Murray and, later, Matthew Flinders had completed only rough surveys – Baudin, by all reports, showed no interest in examining it himself. This is difficult to explain. French charts of Port Philip Bay would not only have been of geo-political import but also useful to later expeditions exploring the south coast. A visit to the bay would also have offered to the voyager-naturalists opportunities for further scientific fieldwork. Perhaps Baudin was influenced to a degree by concern that his investigation of Port Philip Bay would encroach too much on British colonial interests and thus potentially cause conflict. There had certainly been considerable anxiety around the sharing of knowledge, particularly territorial knowledge, between the British and the French. King had been suspicious of Baudin's objectives and Baudin had been at pains to allay those fears. However, had the commander deemed this first-hand knowledge vital for the fulfilment of his mission it is unlikely that he would have been deterred by diplomatic discretion – especially when he was no longer the governor's guest. A more likely reason, if less interesting, is that Baudin judged that this particular addition to the itinerary would be too time-consuming and possibly, depending upon the reports he had received about the bay, too problematic as well – in essence, the result of prioritization. Most of the additions he did make were islands, rather than sections of the mainland coast. He visited King George Sound – not a location originally scheduled – and this sojourn would result in new charts and further fieldwork; however, that was merely a consequence of the arising need for a shore visit. As well as King Island, Kangaroo Island and the 'isles du Romarin', which Baudin mentioned to Jussieu and the Minister of Marine, the expedition also ended up charting the Hunter Islands, in Bass Strait. On the one hand, this focus on islands reflects Baudin's preoccupation with the 'safety of navigation', evident in his correspondence,[9] and with geographic accuracy – it was in relation to these areas that existing charts and surveys were most lacking. On the other, it may also reflect something highlighted by Jean Fornasiero, Peter Monteath and John West-Sooby: the French predilection for insular places, deemed easier to defend.[10]

In any case, it is clear that Baudin was valuing knowledge by its accuracy rather than its breadth. This understanding essentially shaped not only how he designed the itinerary but how he prepared to complete it. In his letter to Jussieu, he wrote:

> I fear that all this work will take longer than the provisions that I have been able to obtain here will permit, because geographic observations demand considerable time; and surveys carried out too quickly will be superficial, imperfect and filled with errors.[11]

He was remembering the first campaign – how shortages of provisions had repeatedly hampered the expedition's work and produced a level of quality in its

documentation that he intended, above all, to avoid repeating. Latour and Bourguet, writing particularly of natural history at the height of the Enlightenment,
explain how an expedition without material evidence of its accomplishments
effectively accomplished nothing at all.[12] Baudin had evidence, but was certain
that, for its lack of depth, it failed to substantiate his voyage. As meaningful
charts required time, and time was provided by a sufficiency of food and water,
one of the tasks that most concerned Baudin at Port Jackson was the gathering of supplies for the next voyage. Over these months, he drew upon money
left by expeditioners who had died during the first campaign – the government
would reimburse those men's heirs[13] – engaged in barter and resorted to bills of
exchange to obtain essential campaign supplies from Commissary John Palmer
as well as from various merchants and farmers throughout the County of Cumberland. He relied heavily upon his skills of negotiation and, while his efforts
were fruitful, the results may not have been adequate without a certain amount
of good luck. Shortly after the expedition set sail, as Baudin explained to the
Minister of Marine,

> We had the good fortune that ... several ships arrived from England, which consider-
> ably diminished the price of salted meat and flour and enabled us to put together a
> year of supplies for the *Casuarina* and for me, the *Naturaliste* has only enough for
> eight months and will not need to call in at any port.[14]

By the time the expedition departed from Port Jackson, it carried more supplies
than it had been provided with at Le Havre.[15]

This would be critical for maintaining the voyager-naturalists' health, discipline and productivity – that is, as long as the provisions were of sufficient
quality to last the months ahead in the damp conditions of the ship. On the
south coast, earlier in 1802, the efficiency of the *Géographe* had steadily deterio-
rated after two casks of salted meat, opened eagerly by a famished and fatigued
crew, were found to be rotten. The biscuit, Baudin had noted with disgust, 'was
not very wonderful either and had begun a long time earlier to crumble into
dust, being riddled with worms and mites'.[16] At Port Jackson then, while preparing the *Naturaliste* to transport the expedition's natural history collection
to France, and the *Géographe* and the *Casuarina* to commence a new scientific
voyage, Baudin refused to compromise on the quality of shipboard provisions.
When several loads of biscuit were found to be mouldy, he ordered that they
be returned to the Sydney bakery and complained angrily to Commissary John
Palmer.[17] And, after another such near disaster, Baudin required that an officer
always accompany crew members to collect the biscuit from the bakery and
inspect it carefully before allowing it to be embarked aboard the ship. He even
resorted to ordering a quantity of biscuit to be delivered by boat from as far away
as Hawkesbury.[18]

With equal diligence, Baudin used the time and exchanges in port also to gather medical information and products. Even with adequate supplies of sound provisions, ships were unhealthy places and, for its effect on both manpower and morale, disease was a constant threat to the functioning of the expedition. Baudin obtained large quantities of a product that would prove vital in preventing scurvy during the upcoming campaign: lime juice.[19] He also exchanged knowledge about new medicines with Matthew Flinders and others in the colony.[20]

All these supplies would not need to stretch as far as those of the initial voyage had done, for Baudin ensured that the new expedition was composed of far fewer men. In Paris, the scientific and naval staff had been chosen by the members of the Institut commissioned to plan and prepare Baudin's voyage. Although a number of the men, midshipmen like Hyacinthe de Bougainville for instance, were considered upon recommendation, the commission members claimed to have 'consulted in their choices only the merit of their learning and of their character perfected by education'. These attributes 'counted for much', they pointed out, 'in a long voyage, the success of which depends upon a perfect harmony between the collaborators of the expedition'.[21] Still, the distance of these savants and politicians from the potential voyagers, particularly the naval men, resulted in the selection of a particularly large group of men, many of whom, though their formal education was no doubt excellent, had little or no experience in natural history expeditions. By the time the *Géographe* and the *Naturaliste* had moored at Bennelong Point, two years later, the situation was rather different: Baudin had the opportunity to make his own choices based on direct observation of the men's contribution to the expedition. Certain changes were made without delay, soon after the *Naturaliste*'s return to the port in late June, but the official staff review did not take place until 4 November – two weeks before the end of the sojourn. This no doubt prevented disruption among the ships' companies while they remained guests of the British colonists. It also gave Baudin more time to evaluate individuals' potential. In the end, thirteen officers and fifty-nine crew members, some for their ill-health and some for their inadequate performance in the expedition, were chosen to return to France with Hamelin.[22] Baudin kept aboard the *Géographe* five officers, including the chief surgeon, François Lharidon, just one midshipman, Charles Baudin, and all of the naturalists excepting only Louis Depuch, too ill to continue the voyage. Aboard the *Casuarina* Baudin transferred Louis Freycinet as captain, Léon Brèvedent as midshipman and fourteen '*élite*' sailors from the *Naturaliste*.[23] The restructuring of the ships' complements was not simply aimed at decreasing numbers and relieving the expedition of its least committed and able members but also at building a team, led according to the stronger style of command Baudin had developed, that might possess the skill, ambition and high standard of discipline required to complete the thoroughgoing work ahead.

And, of course, this was organized according to a new structure, with the introduction of the *Casuarina* in place of the *Naturaliste*. Scholars have commented very little upon this change, typically noting only that the new vessel was smaller and had a shallower draught which allowed closer coastal surveys. However, it was a crucial element of Baudin's strategy for the second campaign. Sorrenson advises that one consider who it was that commissioned a scientific ship,[24] and, in this case, it was the commander himself who not only chose the *Casuarina* in Sydney but who, two years earlier, had chosen the *Géographe* and the *Naturaliste* at Le Havre. As Horner states, he had therefore been to blame for the shortcomings of the consort vessel.[25] It should be noted, though, that Baudin's maritime experience had until then been of naval campaigns and botanical voyages, rather than expeditions of geographical research; moreover, he probably still had in mind the type of voyage he had initially proposed to the Institut – that is, a grand circumnavigation of the globe, with three ships, in the style of the celebrated Enlightenment voyages of Cook, Bougainville and La Pérouse, though more fruitful in its contribution to the natural sciences and their applications.[26] Such a voyage focused on the broad brushstroke of expansive geographic discovery and natural history collecting, rather than the precision of maritime fieldwork envisioned in this case by the commission of the Institut. By the time he stepped into the shipyard in Sydney, following his 'difficult' first exploration of Australia's coastline, Baudin knew precisely what was required to satisfy his instructions. With King's permission, he negotiated the purchase of the yet unfinished *Casuarina* with shipbuilders and merchants James Underwood and Henry Kable. Under the condition that they would 'make her compleat and in the water', he paid 50 points, 230 gallons of rum, 15 bolts of canvas, 16 hundredweight of rope, one ton of bolt and rod iron, and half a barrel of gunpowder.[27] The small ship was launched in August and, saluting the town as she left Sydney Cove with perrior shots answered shot for shot, she took her place beside the other French ships in Neutral Bay.[28] The *Casuarina*, built of native timber and constructed at Port Jackson, would play a key role in the French charting of Australia.

The role Baudin assigned to the new ship was limited and inflexible. Louis Freycinet had accepted the role of captain of the *Casuarina* on the understanding that it would advance his naval career and on the assumption that, as he and his crew were to be given most of the geographical work, Baudin would provide him with 'all the facilities and means possible to carry out this task with exactitude'.[29] He had been delighted at the prospect of playing such a vital part in the expedition.[30] Yet, as he put forward his demands, including even to replace his midshipman, it became clear that he had overestimated the degree of authority and autonomy he had been granted. Baudin explained to him that the *Casuarina*

Would have no other purpose than to enable a more accurate survey to be carried out of the coasts where lack of depth or other difficulties might prevent my ship from getting sufficiently close. You should therefore take particular care never to get far away from the *Géographe*, either by day or by night, except if you have received a particular order to do so. In all circumstances you must handle your ship in such a way that you are never out of sight.[31]

In fact, it was partly in order to ensure that the *Casuarina* remained within his sight throughout the next voyage that Baudin had it fitted out with limited equipment. 'As long as you remain with the *Géographe*', he told the young captain, 'I will provide you with all the assistance necessary'.[32] And, in response to Freycinet's repeated demands, he enquired: 'Am I to understand that you wish to become separated from me?'[33] There is a sense of neuroticism in Baudin's efforts to control and perfect the new campaign; however, beyond an absolute determination to satisfy the aspirations of the French government and scientific community, he had more immediate reasons for such inflexibility. He must have had in mind the long separations between the *Géographe* and the *Naturaliste*, as well as the fact that sub-lieutenant Louis Freycinet was captaining for the first time, and perhaps, therefore, should not claim the same degree of independence as commander Hamelin had enjoyed. In fact, he clearly intended to use this opportunity to give Freycinet training in the command of a scientific ship. In addition to instructions concerning his role in the forthcoming voyage, Baudin also gave him quite detailed advice about how to conduct himself towards his men. This included how to use the spaces of the ship to express the hierarchical structure of the company and his own authority: Brèvedent 'will share your table but be berthed forward, in the cabin constructed for him', he declared.[34] Of course, it was not merely training. The more effectively order was maintained aboard the *Casuarina*, the easier it would be for Baudin to maintain control over the geographical work it carried out by its men – work he perceived as absolutely vital to the success of the expedition. In his final letter to Jussieu before the campaign, he wrote: 'from now on I will be able to examine everything and leave nothing to be done by those who follow after me on similar ventures' – and this would be owed, he trusted, to his implementation of the new ship.[35] Rather than a consort in the usual sense, the *Casuarina* was to be a navigational and charting tool – an extension of the *Géographe*.

This left the *Géographe* itself as the expedition's 'floating laboratory'. As Baudin expressed very clearly to Jussieu, the departure of the *Naturaliste* with the natural history collection marked not the end but a new beginning for the expedition's botanical, zoological, mineralogical and ethnographic fieldwork. The *Géographe* would carry all the remaining naturalists, as well as additional equipment – cages, glass boxes, drawing paper and charcoal – that Baudin had purchased from the colonial stores in Sydney. Accordingly, it would also carry all the animals, plants and objects gathered during the course of the second campaign and provide the space within which the naturalists could collaborate as well as prepare and preserve their findings – essentially, it was set up to ensure the stability and mobility of the natural history collection.

Furthermore, the *Géographe* was to follow an itinerary eminently suited to advancing the expedition's accumulation of knowledge about Australia's natural environment. It would include more frequent landfalls, which would not only provide opportunities for collecting and observing but also allow the astronomer, as he pointed out himself, to maintain the accuracy of the ship's clock – by facilitating navigation, this would in turn maintain the efficiency of the scientific expedition in general. More frequent visits ashore would also be beneficial to the health of all on board. But what was particularly important to Baudin was to allow the naturalists ashore along the south coast, where the *Géographe* had been unable to land during the first campaign. This was the region he had identified in his initial voyage proposal as demanding the attention of voyagers in every field of science.[36] It was also where Fleurieu later instructed him to enable his men to 'penetrate as far inland as possible to ascertain whether this country … offers unknown species of animals and products of interest to botany and mineralogy'.[37] The Flinders expedition had recently sent shore parties inland at several points along this coastline but in effect it remained an untilled field for Western knowledge. Baudin's naturalists would also be allowed ample time to indulge in their research. Even had they been able to reach the shores the first time around, they would have been obliged to conduct their work with all haste, in order, as Fleurieu himself had noted, to avoid the onset of winter.[38] However, the new itinerary was better timed for the seasons; it would place the *Géographe* upon the south coast in the summer.

And in the end, it was in fact the research undertaken on the south coast that was most fruitful. At King George Sound alone, botanist Théodore Leschenault collected around 200 new species of plants and gardener Antoine Guichenot even more.[39] Similarly, in the inventories compiled by François Péron, the provenances most frequently cited were King Island, the Island of St Peter and King George Sound.[40] Yet more significantly, many of the zoological specimens collected in these places by, or at least attributed to, Péron, represented species new to European knowledge, such as the kangaroo from Kangaroo Island.[41] Several are now endangered and no longer inhabit the regions where the Frenchmen collected them, such as the common wombat (*Vombatus ursinus ursinus*),[42] collected at King Island, and the Tamar Wallabies from the Island of St Peter (*Kaguus eugenii*). Some of the species are now extinct, including the King Island dwarf emu (*Dromaius ater*) and the Kangaroo Island dwarf emu (*Dromaius bauinianus*). Great care was taken during the campaign to keep these specimens alive. Baudin obliged officers to give up their cabins to them and he, himself, fed them by hand. Many in fact survived the return voyage and lived on for some time in the menagerie at the Paris Jardin des Plantes and in the garden of Josephine Bonaparte. They were among the expedition's most celebrated contributions to European science.

In fact, the natural history collection accumulated during Baudin's new voyage was exceptional in all. Comprising a total of 200 crates of specimens altogether,[43] which included 100 crates of live plants and 24 herbariums containing 1,500 plant species,[44] 44 crates of zoological specimens and 10 crates of ethnographic objects,[45] it was, according to the director of the Muséum National d'Histoire Naturelle, Antoine-François de Fourcroy, the largest that France had ever received.[46] Indeed, Baudin had outdone himself, for it was even larger than the collection brought to France earlier by the *Naturaliste*. In June 1803, André Thouin, botanist at the Muséum, had reported the arrival at Le Havre of thirty-seven crates of zoological specimens and twelve crates of botanical specimens.[47] Despite Baudin's lengthy instructions on how to care for the plants – how often to water them, where to keep them on deck, how to protect them from sea water and misuse by the sailors[48] – of the 800 live plants placed under Hamelin's care, only 20 had survived the voyage. It was a 'deplorable loss', Thouin declared, and clearly attributable to ill-managed care.[49]

There were two exceptions to the superiority of the fieldwork completed in the second campaign by comparison with that of the first. First of all, the mineralogical collection seems to have been smaller.[50] The months spent by Depuch and Bailly ashore at Port Jackson, where they were able to explore widely, even to the foot of the Blue Mountains, and to receive donations of mineral samples from King and Paterson, must have considerably boosted their collection. And, of course, with Depuch leaving Australian waters aboard the *Naturaliste*, Bailly was the sole mineralogist with the new expedition.

Second, the ethnographical work makes for a more complex comparison. Certainly, the reports were fewer, and more concise, and the number of objects collected was fewer as well. During the second campaign, there were no prolonged or sets of frequent encounters as those previously provided by the visit to Tasmania and its adjacent islands. Yet, there were interesting encounters in places Baudin and his men were visiting for the first time. At King George Sound, Baudin himself, Bailly and geographer Pierre Faure discovered 'two rather peculiar and interesting monuments erected by the natives', some spears left against a tree – which they 'picked up' – as well as series of dikes, constructed 'skilfully and symmetrically' along a river 'in the form of locks'. Baudin took this latter find as 'proof that they are not without intelligence'.[51] And, in the same area, Ransonnet experienced an encounter with a group of five Aboriginal men. It was the Aborigines themselves who initiated contact; in fact, they had invited the French party to approach them the previous day but, for reasons Ransonnet did not explain, it was not then taken up. In any case, Ransonnet and a small group of sailors in the end spent most of the day with these men. In exchange for items he happened to have upon him, an English knife, two bottles and leather buttons he tore from his jacket, Ransonnet obtained from them a spear and spear-thrower, as well as an axe, which he added to the collection aboard the *Géographe*. Afterward, he recorded for Baudin his observations of the men's appearance, speech and song, and noted the care they took to prevent him and the sailors accompanying him from venturing any further inland, where three women – their wives, Ransonnet thought – were keeping a distance.[52] Later, upon landing at Shark Bay, Ronsard found a village of twelve to fifteen huts and, upon Baudin's order, it was reproduced by one of the artists aboard the *Géographe*.[53]

Altogether, these encounters with Indigenous people and belongings considerably developed the expedition's representation of West Australian Aboriginal life, which had previously been rather limited. Yet, these were the only Aboriginal encounters of the second campaign. Neither King Island, Kangaroo Island nor the Island of St Peter was inhabited by Aborigines, and the ships were unable to anchor on either the north-west or north coasts. These circumstances may have been unforeseen by Baudin; still, it is interesting that he made no mention of objectives concerning 'the observations of man' when he described to Jussieu and the Minister of Marine his plan for the new voyage. During these months, Aboriginal life was once more a novelty, a curiosity. The Frenchmen returned to noting smoke from campfires in the distance and examining monuments and empty huts. But the tone of their records suggests they were no longer impatient to make human contact. Baudin's description of the monuments discovered at King George Sound, and in particular of the discussion this discovery prompted – 'everyone tried to guess what these monuments could be and we reasoned in various ways' – is animated. He noted that there seemed to be Aboriginal people 'in the neighbourhood', for he could see columns of smoke further inland, but, without any obvious sign of regret, he concluded that they were too far away to be reached safely. Instead, he left gifts of medals and glass beads and planted some maize and 'other garden seeds' near the monuments – a routine, almost obligatory, gesture.[54]

It was the task of charting that far more deeply absorbed Baudin's attention, as his plans for this voyage indicated it would, and in fact his preoccupation with accurate and thorough geographic work was clearly reflected in the geographical results. During this campaign, the French conducted the first recorded navigation of both King Island and Kangaroo Island and produced the most accurate and thorough charts for the time of Hunter Island, Denial Bay (on the Eyre Peninsula) and the north-west coast of Australia. This work must be attributed mainly to the use of the *Casuarina*, and more specifically, as midshipman Charles Baudin remarked, to the collaboration between Baudin and Louis Freycinet. Neither the *Géographe* nor the *Naturaliste*, on their own, would have been able to sail close enough to these areas with any degree of exactitude. In fact, the *Géographe* only passed safely through the dangerous north-western waters, which for centuries had remained ineffectually charted, because the *Casuarina* guided her, little by little, through the scattered sandbanks and islets.[55] Also, the frequent landfalls enabled topographical studies and opportunities for geographers Faure and Pierre Boullanger to chart islands and bays from small boats – gaining closer proximity to the coastline and achieving greater detail than even the *Casuarina* could have allowed. The geographic work, however, did not always run smoothly. The *Casuarina* was considerably slower than the *Géographe*, and it was her laggard pace that prevented the expedition from reaching the Gulf of Carpentaria. Still, charting the gulf may not have been a critical requirement: in one of his sets of instructions, Fleurieu had instructed Baudin to undertake this task only 'provided that the general examination does not encounter too many difficulties and that [Baudin] does not foresee too great a danger in it'.[56]

In this final leg of the voyage, general fatigue, the commander's deteriorating condition and the impending monsoonal season were all catching up with the expedition, while, to make matters worse, provisions were finally beginning to run low. Still, due to the frequent landfalls and the high stocks of medicines and other supplies, in human terms this campaign, as Jussieu noted, 'was not disastrous as was the first'.[57] Compared to twenty-two deaths over the course of the voyage leading up to the Port Jackson sojourn, during the new voyage the expedition lost only one man in Australian waters. Bernier's death was the result not of unhealthy conditions on the ship but, like so many deaths aboard scientific and discovery vessels, of disease contracted at Timor. Similarly, relations on board had been smoother than during the first campaign. There were disagreements between Baudin and Louis Freycinet on the south coast, particularly when they missed a rendezvous at Kangaroo Island, but on the whole and particularly on the west coast, they collaborated well. Baudin managed to avoid delegating authority to a second in command until the stopover at Timor, by which point most of the work had been completed. In port, Henri Freycinet and Ronsard came into conflict over the question of seniority – their feelings clearly had not changed since Port Jackson – and Baudin finally yielded and put the decision to a vote. The tension between the lieutenants, and probably between them and Baudin, must have been rising during the course of the voyage, but it is not obvious in the records of those months and it would not appear to have hindered the expedition's work. In fact, the second campaign's productivity must owe much to the soundness of these human elements, which enabled the scientific ships to function efficiently.

The second campaign had effectively been completed when the *Géographe* and the *Casuarina* weighed anchor in Kupang Bay, Timor, in May 1803. There, in a letter to the Minister of Marine, Baudin reported on the work achieved by the expedition since the date of his last letter, as he had done at each port-of-call. However, as this time he brought the Australian voyage to a close, he made a stronger and more direct attempt to attract the government's attention to his accomplishments. He commenced his closing comments: 'I end this letter, Citizen Minister, by asking you to recall me to your Memory and to those of the Consuls should the occasion present itself', and continued:

> I hope the Government will be pleased with us when it learns of all that we have done to this point and despite the small Imperfections that may be found in the results of our work this voyage will be no less useful for the future than honourable for those who have undertaken it.

It was in the final lines that he hinted at something more specific and significant in the expedition's results:

> I will not enter into further detail as this letter might pass through several hands before reaching you and it would not be appropriate if it did not reach you intact given the importance of what I have to communicate to you.[58]

What had the commander to report, that was not for the eyes or ears of anyone but the Minister himself? One can only speculate, for no further words from Baudin on this subject have yet come to light, however it is most likely that the message related to Port Jackson and the British. Alternatively, it is also possible that, despite the attitude he had demonstrated earlier in his letters to King, it entailed a considered proposal for a French settlement somewhere on the Australian coast – Geographe Bay perhaps – or at d'Entrecasteaux Channel, Tasmania.

Baudin was a politically astute man. He clearly recognized that not only was scientific research tied up in national concerns and even political affairs – via its applications, its influence on international relations of knowledge and prestige and the roles of the savants themselves – but that the development of his career in scientific voyaging depended upon his success in satisfying the imperial interests of the French Consulate. It was therefore not only with the fundamental and explicit elements of his instructions in mind but also with an acute awareness of the underlying geo-political imperatives of this mission that he redesigned and rebuilt the expedition at Port Jackson. And sharpening all these considerations was, most significantly, the experiences and knowledge he had obtained in his role as French envoy and scientific captain *sur la terrain*. He prepared the new voyage not simply by recalling his instructions but, with this immediate understanding of the environment to be examined and of the existing claims upon it, according to his own decisions about the knowledge France required and how this knowledge could best be gained. In this way, in fact, he designed a voyage that ultimately functioned, even more closely than had the first, in accordance with the more circumscribed and systematic style of voyaging envisioned by the Institut.

8 EPILOGUE: VOYAGING INTO THE NINETEENTH CENTURY

Baudin composed his last letters at Mauritius, just shortly before he finally succumbed to tuberculosis. 'I have enough strength left at the moment', he wrote to the Minister of Marine, 'to assure you that the intentions of the Government have been fulfilled and that this voyage will be honourable for the French'.[1] The *Géographe* drew into the port at Lorient six months later and, while the naturalists of the Institut were delighted with the collections it brought them, the government paid little attention to its arrival and to the accomplishments of the expedition in general. Moreover, in the long term, despite the vast size of the collection and the meticulous detail and copious extent of the records, Baudin's accumulation of knowledge did not stick in French history, as certain voyages of the eighteenth century had done. Certainly, the expedition's legacy would have been strengthened had the commander, who had planned and conceptualized that second and most successful campaign himself, been able to present the results in person. Yet Baudin's expedition was not unique in returning without its captain, and, in any case, the Consuls and ministers had little time for appreciating distant achievements of any sort on what was the eve of the First Empire. In fact, just as the nation advanced into a new era, so too did scientific voyaging – and maritime expeditions lost their prominence in the history of France.

Over the course of the nineteenth century, scientific ships functioned with increasing pragmatism and efficiency as tools of science and the state. No further maritime expeditions were sent out by Napoleon, whose attention for the remainder of his reign was fixed on reforming France and building it into a continental empire; and while the more peaceful if politically unstable eras of the restored Bourbons and the Orleanists, which followed, were eras of unprecedentedly frequent voyaging to Oceania, it was voyaging that was also more politically and commercially motivated than previously – voyaging still driven at times by questions of science but at other times, first and foremost, by interests in trade and territory. This was not simply because of a perception that there was little left to 'discover': France's final contribution to the Age of Exploration was an ambitious scientific venture led by Dumont d'Urville in 1837–40, with two ships, a hydrographer and a phrenologist, and a comprehensive itinerary that urged d'Urville to better Cook's attempt on Antarctica. However, to the extent that French activities in the Pacific were encouraged by rivalry with the British, French expeditions overall were motivated as much by the example of the flourishing colonial project in Australia as by the accomplishments of Cook.[2] Large expeditions of over 450 men developed trade relationships, reconnoitred potential sites for colonization and asserted national prestige, while the truly scientific expeditions kept matters of imperial policy generally in mind. Following the Baudin expedition's sojourn at Port Jackson, for example, the British colony became a central port-of-call for European voyagers and each expedition of the Restoration era used its time there not only to obtain respite and replenishment but to monitor the settlers' efforts to 'civilize' Aboriginal people and to critique the strengths and weaknesses of the transportation system.

In their scientific objectives, these latter expeditions were concentrated on greater accuracy and greater depth of knowledge on more defined areas of research than previous French expeditions had been – particularly those that preceded Baudin's voyage. Their itineraries were usually extensive, but focused particularly on the least known coastlines and regions, such as New Guinea and its islands, and the expeditions themselves were physically smaller than their predecessors, including the Baudin expedition in its original state. D'Urville's second expedition, with two ships and over 160 men, was an exception; Louis Freycinet, the first scientific voyager to return to Oceania after Napoleon's fall, captained one ship and 125 men, the subsequent expeditions of Duperrey and D'Urville each comprised only one vessel as well as less than 80 men. Moreover, the fieldwork carried out by these expeditions was more systematic and disciplined than that of their predecessors – more uniformly so than that of the Baudin expedition. From Louis Freycinet's expedition onward, and with the exception usually of an artist, they carried only naval staff – the scientific fieldwork was carried out by surgeon-naturalists, who, subject to naval regulations, worked under the direct command of the captain. Like previous voyager-naturalists, they did load their ships with vast collections of plants, animals, minerals and ethnographic objects but were not set on the vague pursuit of encyclopaedic knowledge. While applying recent scientific methods and theories, they concentrated, above all, on gathering data to feed the research interests of sedentary naturalists – questions about terrestrial magnetism and meteorology for instance, and, progressively, the validity and significance of 'racial' difference.

In the end, the data they provided would advance theories of 'race', which served to justify the imperialism of the later nineteenth century, and it would lead in part to the annexation of Tahiti and the establishment of a penal colony in New Caledonia. It would also contribute to the development of countless scientific discoveries in the *cabinets* of scientists in Paris. However, the scientific captains of the post-Napoleonic era did not, like Bougainville after his visit at the 'isle of Venus', provide stories of idyllic lands and cultures, they did not bring home human curiosities, as Bougainville brought Ahu-Toru and Cook, Mai, nor did any of them disappear and become legend, like the La Pérouse expedition, or carry the romance of being in search of a lost expedition, like the voyage of d'Entrecasteaux. Certainly, Dumont d'Urville acquired some notoriety for reaching the magnetic South Pole and establishing the racial/geographic demarcation of Oceania; however, just like Baudin's Australian voyage, these expeditions were fated overall by their very nature to make significant contributions to French history and yet not to occupy a conspicuous place within it.

The scientific voyages of Louis Freycinet, Dumont d'Urville and their contemporaries represented a clear progression from the '*directe*' style of exploration that the commission of the Institut had tested under Baudin. Their itineraries, though more extensive, remained circumscribed, according to yet more targeted objectives. Moreover, discipline and efficiency remained key, and were achieved by precise goals combined with still more pared down expeditions – most notably, via the exclusion of citizen naturalists. In 1800, the commission had assumed that, with naturalists educated in various fields of natural science, Baudin's expedition would complete more specialized fieldwork than that achieved previously; and, indeed, it did. However, as the roles of sedentary and field naturalists were further defined in the first half of the nineteenth century, specialization was increasingly contained to the *cabinet* while broad knowledge and, above all, discipline were demanded from voyagers sent into the field.

In the nineteenth-century field, moreover, natural-science methods and points of view that were just emerging and often uncertainly applied by Baudin's men, were manifested in their more advanced states. The usefulness of the dynameter was questioned by Freycinet in the 1820s; however, the basic reasoning behind Péron's experimentation with tests of physical strength in relation to comparative degrees of 'civilization', and particularly the theory proposed in Cuvier's *Note instructive*, were later to underlie the collection of comprehensive anthropometric data and the bold assertion of significant, perhaps even innate, differences in human characters and capacities. Furthermore, as a result of the accumulation of knowledge acquired first-hand from multiple expeditions and indirectly from the proliferation of published accounts about the region, and also due to the European imprint left on beaches frequently visited by naturalists and missionaries, French voyagers developed a familiarity with the environments and peoples of Oceania – a familiarity similar to that felt by Baudin and his men at Port Jackson, indeed, a familiarity which brought island scenes squarely beneath a colonial lens.

Scientific voyaging was constantly evolving, but Baudin's expedition was particularly distinct from those which had preceded it in the eighteenth century as well as those which followed it in the decades afterward. More precisely, it possessed some similarities with each group: on the one hand, the naturalists', particularly Baudin's, relatively indiscriminate collecting, as well as their lingering, hopeful, faith in the concept of the noble savage, were fairly typical of Enlightenment thought and practice, while, on the other hand, the attempt to render it a more systematic, disciplined, venture, and the emergence of 'scientific' outlooks and methods, anticipated succeeding expeditions. However, it also possessed certain characteristics unique among French expeditions: most notably, its large number of scientific staff and its concentration upon the potentially single continent of 'Terra Australis'. It constituted a transitional moment in scientific voyaging – a moment produced from a period of particularly rapid change in the world of the Institut and the Muséum and, more broadly, in society under Bonaparte.

This was elucidated most clearly during that gap between voyages that was created when Baudin set about constructing a new expedition at Port Jackson. In these months, cross-cultural contacts, shipboard relations and scientific fieldwork were conducted within a liminal paradigm at once more liberating and more confronting than that which operated normally at sea – an ambiguous paradigm that brought particular tensions to the surface. While certain of the naturalists and officers, including the commander himself, would not have been a part of this expedition had it occurred much earlier in French history, others took their careers for granted. Certainly, the time for political insurgence, like that which disrupted d'Entrecasteaux's Revolutionary voyage, had passed; instead, the inter-relationships among Baudin's men were heavy with contained hostilities and, among those embarking on the new voyage, intense ambition. In the fieldwork carried out by Baudin and the naturalists, tensions around collecting, reporting, analysing and reflecting were evident – their boundaries were unclear. Altogether, the voyagers' identity as Frenchmen of the new Consulate era was strengthened in definition against the English colonists and, at least for some, simultaneously brought into question via observations of Port Jackson's Aboriginal people. Indeed, there was a tension, felt certainly by Baudin and to varying degrees by many of his men as well, between the theories, methods and underlying principles they were expected to apply in the field and their feelings about the fundamental humanity and the richness of the environment that confronted them. During their time at Port Jackson, there emerged a self-consciousness and ambivalence about European intrusion in this Indigenous world that usually, when either of the voyages were in progress, was quelled by the conceptual distance and silence between themselves and the beach. Increasingly, during the course of subsequent Oceanic expeditions, the imperatives of anthropological fieldwork would more convincingly prevail over any recognition of a shared humanity.

In Sydney, in that unique temporal and colonial space, the commander, in particular, had reengaged with the contemporary imperial and scientific cultures of end-of-the-Revolution France and reconsidered his expedition's purpose in the light of new experiences and new understandings. While, initially, in Paris, he had proposed a grand global venture in natural history voyaging, he now realized with renewed clarity that it was not another Bougainville, La Pérouse or even d'Entrecasteaux that was required. Like their voyages, certainly, his expedition might boost national prestige; however, the actual work that it was to accomplish and how accordingly it was intended to function, were rather different.

Indeed, Baudin's expedition was not truly the 'last of the great adventures'.[3] It was an outcome of the Age of Enlightenment, to an extent, but it also marked, by design, a major turn in scientific voyaging. Above all, it was a product of this immediately post-Revolutionary moment – the early era of the Consulate. It thus captures the historic effort of French naturalists to reorganize and reframe their investigations of the natural world; it captures individual experiences of new democratic principles and opportunities in society, in naval service and in discovery; it captures the evolving nature of scientific voyaging; and, finally, it also captures a moment in his leadership of France when Bonaparte most overtly engaged with Enlightenment values and Revolutionary ideas, when he was willing to offer his support to exploration in the South Seas, a moment just prior to Empire.

NOTES

Introduction: Voyaging out of the Enlightenment

1. 'un voyage autour du monde ... qui intéresse l'Europe entière', Nicolas Baudin to the members of the Institut National, 6 floréal an VIII [26 April 1800], Archives Nationales de France, Série Marine (hereafter ANF, SM), BB4995. See also Nicolas Baudin to the Minister of Marine and the Colonies, undated, ANF, SM, BB4995.

2. 'pour ajouter aux découvertes des grands navigateurs qui depuis 40 ans a singulièrement aggrandé la domaine de ses deux sciences'. 'Rapport sur le voyage entrepris par les ordres du gouvernement et sous la direction de l'institut, par le Capitaine Baudin', 26 December 1800, Muséum Nationale d'Histoire Naturelle, Paris (hereafter MNHN), ms 1214/6.

3. 'engagerai le gouvernement ami des sciences'; 'déterminer les consuls à faire exécuter ce que les autres avaient négligé'. 'Rapport sur le voyage', MNHN, ms 1214/6 and N. Baudin, *Mon voyage aux terres australes: journal personnel du commandant Baudin, illustré par Lesueur et Petit*, ed. J. Bonnemains (Paris: Imprimérie Nationale, 2000), p. 30.

4. 'La réception fut telle que je l'avais prévue, le voyage décidé'. Baudin, *Mon voyage*, p. 5.

5. 'on donna la préférence à des voyages circonscrits à des points déterminé, dirigés vers des côtes moins connues ... vers les nouveaux archipels dont il reste à vérifier le nombre, l'étendue, les contours et la population des îles qui en sont partie. Ces expeditions directe' 'Rapport sur le voyage', MNHN, ms 1214/6.

6. J. Dunmore, *French Explorers in the Pacific, vol. 2: The Nineteenth Century* (Oxford: Oxford University Press, 1969), p. 11.

7. See for instance: G. Stocking, 'French Anthropology in 1800', *Isis*, 55:180 (1964), pp. 134–50; M. J. Hughes, 'Philosphical Travellers at the Ends of the Earth: Baudin, Péron and the Tasmanians', in R. W. Home (ed.), *Australian Science in the Making* (Cambridge: Cambridge University Press, 1998), pp. 23–44; J. Fornasiero and J. West-Sooby, 'Taming the Unknown: The Representation of Terra Australis by the Baudin Expedition, 1801–1803', in A. Chittleborough, G. Dooley, B. Glover and R. Hosking, *Alas for the Pelicans: Flinders, Baudin and Beyond* (Kent Town: Wakefield Press, 2002), pp. 59–80; M. Sankey, 'The Aborigines of Port Jackson, as seen by the Baudin Expedition', *Australian Journal of French Studies*, 41:2 (2004). pp. 117–51; M. Jangoux, 'Les zoologistes et botanistes qui accompagnèrent le capitaine Baudin aux terres australes', *Australian Journal of French Studies*, 41:2 (2004), pp. 55–78; J.-L. Chappey, *La société des observateurs de l'homme: des anthropologues au temps de Bonaparte* (Paris: Société des études robespierristes, 2002), pp. 465–7.

8.   See for instance: J. Fornasiero, P. Monteath and J. West-Sooby, *Encountering Terra Australis: The Australian Voyages of Nicolas Baudin and Matthew Flinders* (Kent Town: Wakefield Press, 2004); S. Konishi, *The Aboriginal Male in the Enlightenment World* (London: Pickering & Chatto, 2012); M. Sankey, 'The Baudin Expedition in Port Jackson, 1802: Cultural Encounters and Enlightenment Politics', *Explorations*, 31 (December 2001), pp. 5–36; Chappey, *La société des observateurs de l'homme*, p. 164.

9.   J. Gascoigne, *The Enlightenment and the Origins of European Australia* (Cambridge: Cambridge University Press, 2002), p. 1.

10. See for instance, D. Outram, *The Enlightenment*, 2nd edn (Cambridge: Cambridge University Press, 2005), pp. 126–40.

11. Ibid., pp. 52–3.

12. Dunmore, *French Explorers in the Pacific*, vol. 2, p. 384.

13. Outram, *The Enlightenment*, pp. 94–5.

14. 'Plan of Itinerary for Citizen Baudin', in N. Baudin, *The Journal of Post-Captain Nicolas Baudin, Commander in Chief of the Corvettes* Géographe *and* Naturaliste, *Assigned by Order of the Government to a Voyage of Discovery*, trans. C. Cornell (Adelaide: Libraries Board of South Australia, 1974), p. 1.

15. C. E. Harrison, 'Projections of the Revolutionary Nation: French Expeditions in the Pacific, 1791–1803', *Osiris*, 24 (2009), pp. 33–54, on p. 35 and medal commemorating the voyage of discovery commanded by Nicolas Baudin, 1800, State Library of New South Wales, R 942.

16. Harrison, 'Projections of the Revolutionary Nation', p. 34.

17. Ibid., p. 36.

18. Ibid., p. 35.

19. C. E. Harrison, 'Replotting the Ethnographic Romance: Revolutionary Frenchmen in the Pacific, 1769–1804', *Journal of the History of Sexuality*, 21:1 (January 2012), pp. 39–59, on p. 40.

20. F. Horner, *Looking for La Pérouse: D'Entrecasteaux in Australia and the South Pacific, 1792–1793* (Melbourne: Melbourne University Press, 1995), pp. 267–8.

21. See for instance Fornasiero and West-Sooby, 'Taming the Unknown', pp. 59–80; Jangoux, 'Les zoologistes et botanistes qui accompagnèrent le capitaine Baudin' and N. Starbuck, 'The Colonial Field: Science, Sydney and the Baudin Expedition (1802)', *Explorations*, 52 (June 2012), pp. 3–35.

22. Certain historians describe Bonaparte as in fact heavily influenced by Enlightenment philosophy, for example G. Rudé, *Revolutionary Europe, 1783–1815* (Glasgow: Fontana, 1964), p. 223–4 and J. Godechot, B. F. Hyslop and D. L. Dowd, *The Napoleonic Era in Europe* (New York: Holt, Rinehart and Winston, 1971), pp. 33–4. Many others comment that Bonaparte deliberately represented himself as an intellectual in order to gain the support of the *ideologues* in the lead-up to the coup of brumaire and into the initial period of his leadership as First Consul, for example: J. Tulard, *Napoleon: The Myth of the Saviour*, trans. T. Waugh (London: Methuen and Co., 1985), p. 67; C. Jones, *The Great Nation: France from Louis XI to Napoleon* (London: Penguin, 2003), p. 578 and P. Dwyer, *Napoleon: The Path to Power, 1769–1799* (London: Bloomsbury, 2007), pp. 345–7 and 477–9.

23. That Bonaparte recognized the strategic advantages of the Baudin expedition but did not intend it to carry out any territorial or military objectives has repeatedly been demonstrated, however, it is particularly well argued by Dunmore, *French Explorers in the Pacific*, vol. 2, p. 9.

24. See P. McPhee, *A Social History of France, 1789–1914* (Basingstoke: Palgrave Macmillan, 2004), p. 86 and Dwyer, *Napoleon*, p. 338.
25. S. Wolf, 'French Civilization and Ethnicity in the Napoleonic Empire', *Past and Present*, 124 (August 1989), pp. 96–120, on p. 103.
26. See for instance: M.-N. Bourguet, 'Race et folklore: L'image officielle de la France en 1800', *Annales. Histoire, Sciences Sociales*, 31:4 (July–August 1976), pp. 802–23; Wolf, 'French Civilization and Ethnicity', pp. 96–120; D. Outram, 'New Spaces in Natural History', in N. Jardine, J. A. Secord and E. C. Spary (eds), *Cultures of Natural History* (Cambridge: Cambridge University Press, 1996), pp. 249–65; C. Blanckaert, '1800 – Le moment "naturaliste" des sciences de l'homme', *Revue d'Histoire des Sciences Humaines*, 3 (2000), pp. 117–60; Chappey, *La société des observateurs de l'homme*, pp. 225–380; Harrison, 'Projections of the Revolutionary Nation', pp. 33–52 and 'Replotting the Ethnographic Romance'.
27. Blanckaert, '1800 – Le moment "naturaliste"', p. 118.
28. See ibid., pp. 117–60.
29. See G. Dening, *Performances* (Melbourne: Melbourne University Press, 1996), pp. 108–9.
30. Dorinda Outram discusses the significance of spaces in natural history research during this era and notes that the concept of the 'field' has not yet been defined: Outram, 'New Spaces in Natural History', p. 259. For my own discussion about the nature of natural history fields and specifially the 'colonial' field, see Starbuck, 'The Colonial Field'.
31. For example, Susan Hunt and Paul Carter declare that 'King received [Baudin] and his men, nursing, feeding and entertaining them in the next five months with an extraordinary generosity'. S. Hunt and P. Carter, *Terre Napoléon: Australia through French Eyes, 1800–1804* (Sydney: Historic Houses Trust of New South Wales, 1999), p. 23.
32. N. Thomas, *In Oceania: Visions, Artifacts, Histories* (London: Duke University Press, 1997), pp. 19 and 41.
33. G. de Beer, *The Sciences Were Never at War* (London: Thomas Nelson and Sons, 1960). For an examination of this concept in relation to the Baudin expedition, see N. Starbuck, 'Sir Joseph Banks and the Baudin Expedition: Exploring the Politics of the Republic of Letters', in G. Betros (ed.), *French History and Civilization: Papers from the George Rudé Seminar*, vol. 3 (2009), pp. 56–68.

1 Between Revolution and Empire: France and its Australian Voyage in 1800

1. Jones, *The Great Nation*, p. 577.
2. See Dwyer, *Napoleon*, pp. 518–19.
3. Ibid., p. 322.
4. Chappey, *La société des observateurs de l'homme*, p. 83.
5. McPhee, *A Social History of France*, p. 79.
6. Ibid., p. 82.
7. Rudé, *Revolutionary Europe*, p. 226.
8. M. Lyons, *Napoleon Bonaparte and the Legacy of the French Revolution* (New York: St Martin's Press, 1994), pp. 50–1.
9. D. Garrioch, *The Making of Revolutionary Paris* (Berkeley, CA: University of California Press, 2002), pp. 305–7. Garrioch is referring here specifically to Paris but, as he

points out himself, these developments affected other parts of France as well, if to vary-ing degrees and at slightly different times. See p. 7.

10. Garrioch, *The Making of Revolutionary Paris*, pp. 307–9 and C. Jones, *Paris: Biography of a City* (London: Penguin, 2004), pp. 285–7.

11. Garrioch, *The Making of Revolutionary Paris*, p. 309 and Jones, *Paris*, pp. 285–7.

12. Quoted in Lyons, *Napoleon Bonaparte*, p. 85.

13. Ibid., p. 109.

14. Quoted in ibid., p. 105.

15. O. Hufton, *Women and the Limits of Citizenship in the French Revolution* (Toronto: Uni-versity of Toronto Press, 1999), pp. 4–5.

16. Lyons, *Napoleon Bonaparte*, pp. 154–5.

17. Frédéric de Bissy to the Minister of Marine and the Colonies, undated. ANF, SM, BB997.

18. M. J. Hughes, 'Making Frenchmen into Warriors: Martial Masculinity in Napoleonic France', in C. E. Forth and B. Taithe (eds), *French Masculinities: History, Culture and Politics* (Basingstoke: Palgrave Macmillan, 2007), pp. 51–66, on pp. 53–4.

19. 'des hommes courageux qui vont affronter tant de dangers pour multiplier nos con-naissances!' L.-F. Jauffret, 'Discours de L.-F. Jauffret à la Société des Observateurs', in G. Hervé, 'Les Instructions anthropologiques de G. Cuvier', *Revue de l'Ecole d'anthropologie de Paris* (Paris: Alcan, 1910), pp. 289–306, quoted in Chappey, *La société des observa-teurs de l'homme*, p. 162.

20. 'List of the Crew of the Corvette *Géographe*' and 'List of the Crew of the Corvette *Natu-raliste*', in Baudin, *The Journal*, pp. 579–85.

21. McPhee, *A Social History of France*, p. 14.

22. W. S. Cormack, *Revolution and Political Conflict in the French Navy, 1789–1794* (Cam-bridge: Cambridge University Press, 1995), pp. 267–75.

23. 'La transition entre l'Ancien Régime des sciences et la révolution intellectuelle'. C. Blanckaert, 'Naissance et Développement d'une Institution: Introduction', in C. Blanc-kaert, C. Cohen, P. Corsi and J.-L. Fischer (eds), *Le Muséum au premier siècle de son histoire* (Paris: Éditions du Muséum National d'Histoire Naturelle, 1997), pp. 19–24, on p. 21.

24. R. Hahn, 'Du Jardin du Roi au Muséum: les carrières de Fourcroy et de Lacepède', in *Le Muséum*, pp. 31–41.

25. Quoted in Outram, 'New Spaces in Natural History', p. 261.

26. Ibid., p. 259.

27. Chappey, *La société des observateurs de l'homme*, p. 250.

28. Blanckaert, '1800: Le moment "naturaliste"'.

29. Harrison, 'Replotting the Ethnographic Romance', p. 40; Bourguet, 'Race et folklore', pp. 811–2, 815, 817; Blanckaert, '1800: Le moment "naturaliste"', p. 135.

30. J.-M. Degérando, 'Considérations sur les diverses méthodes à suivre dans l'observation des peuples sauvages' and G. Cuvier, 'Note Instructive Sur les Recherches à faire relative-ment aux différences anatomiques des diverses races d'hommes, 1800', in J. Copans and J. Jamin (eds), *Aux origines de l'anthropologie française: Les mémoires de la société des obser-vateurs de l'homme en l'an VIII* (Paris: Le Sycomore, 1978), pp. 127–69 and pp. 171–6.

31. M. Staum, *Minerva's Message: Stabilizing the French Revolution* (Montreal: McGill-Queen's Press, 1996), pp. 155 and 171.

32. L.-F. Jauffret, 'Introduction aux mémoires de la société des observateurs de l'homme', in Copans and Jamin (eds), *Aux origines de l'anthropologie*, pp. 73–85, on p. 77.

33. M. Jacob, *Living the Enlightenment: Freemasonry and Politics in Eighteenth-Century Europe* (Oxford: Oxford University Press, 1991), pp. 203–14.

34. Quoted in Staum, *Minerva's Message*, p. 37.

35. Baudin, *Mon voyage*, p. 30.

36. Ibid., p. 31.

37. 'Plan of Itinerary for Citizen Baudin', pp. 74–80.

38. Gazette Nationale ou Le Moniteur Universel du 29 Vendémiaire an IX [21 October 1800], Collection Lesueur, Muséum d'Histoire Naturelle, Le Havre (hereafter CL, MHN Le Havre), dossier 06 002.

39. Gazette Nationale du 29 Vendémiaire an IX [21 October 1800], CL, MHN Le Havre, dossier 06 002.

40. 'De longs et périlleux travaux, Seront bientôt notre partage, Pour les tenter bravons les flots, Bravons les tempête et l'orage; Surtout soyons toujours unis, Et puisqu'un ami nous rassemble, En son honneur, nos chers amis, Rions, chantons, buvons ensemble'. L. Depuch, '(Depuch) poème, chanson en l'honneur du départ de l'expédition avec quelques commentaires de Péron', CL, MHN Le Havre, dossier 06 009.

41. Baudin, *The Journal*, p. 400.

42. Ibid., p. 401.

43. Ibid., p. 402.

44. Nicolas Baudin to the Minister of Marine and the Colonies, Port Jackson, 20 brumaire an XI [11 November 1802], ANF, SM, BB4995.

2 'I Should Wish … to Establish a Few Tents on Shore': The Port Jackson Stay

1. See Journal tenu a bord de la Corvette le Naturaliste par J. Vᵉ Couture Aspirant de 1ᵉʳᵉ Classe, ANF, SM, 5JJ57, entry dated 4–5 floréal an X [24–5 April 1802] and Journal [Nautique] du Naturaliste Giraud aspirans de la marine pendant les années 9. 10 et 11, ANF, SM, 5JJ57, entry dated 4–5 floréal an X [24–5 April 1802].

2. Jacques Félix Emmanuel Hamelin des Essarts, Capitaine de Frégate Commandant le Naturaliste, Corvette destinée avec celle du Géographe sous les ordres du Capitaine Nicolas Baudin pour une expédition de découvertes, armée et equipée pour 4 ans, vol. 2, ANF, SM, 5JJ42, entry dated 5–6 floréal an X [25–6 April 1802].

3. Letter from Emmanuel Hamelin to Philip Gidley King, written at Port Jackson and dated 25 April 1802, in F. M. Bladen (ed.), *Historical Records of New South Wales*, vol. 4 (Sydney: Government Printer, 1896) (hereafter *HRNSW*), p. 942. It has been suggested that Hamelin feared his requests would be refused, see J. Webb, *George Caley: Nineteenth Century Naturalist* (Chipping Norton: Surrey Beatty and Sons, 1995), p. 41. François Péron also claimed that, because of the hatred the British felt for France as a result of the Napoleonic Wars, Hamelin had reason to fear that the colonists at Port Jackson would refuse to assist the expedition. See F. Péron [L. Freycinet], *Voyage of Discovery to the Southern Lands*, trans. C. Cornell, vol. 1, 2nd edn (1824) (Adelaide: Friends of the State Library of South Australia, 2006), p. 290.

4. *Journal de Breton, aspirant de 1ᵉʳᵉ Classe à bord du Géographe: Depuis Le 9 Vendémiaire an 9 Jusqu'au 9. Brumaire an 10*, ANF, SM, 5JJ57, entry dated 5 floréal an X [25 April 1802].

5. 'Il regardait en nous, disait-il, les citoyens de tout le monde et nous avions le droit de nous attendre à la reconnaissance de toutes les nations', *Journal de S' Cricq, Enseigne de Vaisseau sur la Corvette Le* Naturaliste, *commandée par le Citoyen Hamelin Cap.ⁿᵉ de fregᵗᵉ Voyage de découvertes du Cap.ⁿᵉ Baudin*, ANF, SM, 5JJ48, entry dated 5 floréal an X [25 April 1802].

6. P. G. King, 'Regulations to be Observed by the French Ship during her Stay in Port Jackson', dated 27 April 1802, in Bladen (ed.), *HRNSW*, vol. 4, p. 943 and 'Regulations to be Observed by all Masters or Commanders of British or Foreign Vessels Arriving at Port Jackson, and by all Merchants, Importers, and Consignees Resident in His Majesty's territory of New South Wales', undated, in Bladen (ed.), *HRNSW*, vol. 4, pp. 144–6.

7. Saint-Cricq, *Journal*, entry dated 5 floréal an X [25 April 1802].

8. Philip Gidley King to the Duke of Portland, Sydney, 21 May 1802, in Bladen (ed.), *HRNSW*, vol. 4, pp. 761–4, on pp. 762–3.

9. Philip Gidley King to the Duke of Portland, Sydney, 21 May 1802, in Bladen (ed.), *HRNSW*, vol. 4, p. 763.

10. King had been planning to increase Britain's possessions in the region for some time but the government had been reluctant. The risk of a neighbouring French colony must have struck King as a strong justification for his plan. For examples of King's claims about the intentions of the French, which he voiced for the first time following the departure of the *Naturaliste*, see Philip Gidley King to the Duke of Portland, Sydney, 21 May 1802, in Bladen (ed.), *HRNSW*, vol. 4, pp. 764–6, on p. 766 and Philip Gidley King to Joseph Banks, Sydney, 5 June 1802, in Bladen (ed.), *HRNSW*, vol. 4, pp. 782–6, on p. 785. The fact that the letter of 21 May 1802 was the first in which King supported his proposal for a settlement at Port Philip by mentioning the probability of the French planning a settlement on the north-west coast of Bass Strait has also been observed by Manning Clark. See C. M. H. Clark, *A History of Australia*, vol. 1, *From the Earliest Times to the Age of Macquarie* (Melbourne: Melboure University Press, 1962), p. 182. And, for a more comprehensive treatment of the British view of the French expedition and of British fears of French intentions hastening but not instigating the colonization of Port Philip Bay and Tasmania, see J. West-Sooby, 'Une expédition sous haute surveillance: le voyage aux terres australes vu par les Anglais', in M. Jangoux (ed.), *Portés par l'air du temps: les voyages du Capitaine Baudin*, special number of *Études sur le 18ᵉᵐᵉ siècle*, 38 (2010), pp. 187–201.

11. Hamelin, *Journal*, vol. 2, ANF, SM, 5JJ42, entry dated 9–10 floréal an X [29–30 April 1802].

12. Ibid., entries dated 11–12 and 12–13 floréal an X [1–2 and 2–3 May 1802].

13. Breton, *Journal*, vol. 2, ANF, SM, 5JJ57, entry dated 9 floréal an X [29 April 1802].

14. Breton, *Journal*, vol. 2, entry dated 11 floréal an X [1 May 1802] and Hamelin, *Journal*, vol. 2, ANF, SM, 5JJ42, 11–12 floréal an X [1–2 May 1802]; Couture, *Journal*, ANF, SM, 5JJ57, entry dated 11–12 floréal an X [1–2 May 1802]; Hamelin, 'Compte de mes dépenses au Port Jackson pour la Corvette le *Naturaliste*', ANF, SM, BB4997.

15. Hamelin, *Journal*, vol. 2, ANF, SM, 5JJ42, entry dated 12–13 floréal an X [2–3 May 1802] and Couture, *Journal*, entry dated 12–13 floréal an X [2–3 May 1802].

16. Couture, *Journal*, entry dated 21–22 floréal an X [11 May 1802].

17. 'elle m'est étrangère ...; je ne prétends pas faire de l'esprit ni jouer le savant', Hamelin, *Journal*, vol. 1, ANF, SM, 5JJ41, entry dated 13–14 prairial an IX [2 June 1801].

18. Philip Gidley King to Joseph Banks, Sydney, 5 June 1802, in Bladen (ed.), *HRNSW*, vol. 4, p. 782.

19. Hamelin, *Journal*, vol. 2, ANF, SM, 5JJ42, entry dated 19–20 floréal an X [9–10 May 1802] and Saint-Cricq, *Journal*, entry dated 5 floréal an X [25 April 1802].

20. Hamelin, *Journal*, vol. 2, ANF, SM, 5JJ42, entry dated 21–22 floréal an X [11–12 May 1802].

21. Emmanuel Hamelin to the Minister of Marine and the Colonies, Port Jackson, 22 floréal an X [12 May 1802], reproduced in Hamelin, *Journal*, vol. 2, ANF, SM, 5JJ42, entry dated 21–22 floréal an X [11–12 May 1802].

22. F. Horner, *The French Reconnaissance: Baudin in Australia, 1801–1803* (Carlton: Melbourne University Press, 1987), p. 248.

23. Nicolas Baudin to Antoine-Laurent de Jussieu, Port Jackson, 20 brumaire an X [11 November 1802], Muséum National d'Histoire Naturelle (hereafter MNHN), ms 2082, part 8.

24. M. Jangoux, 'La première relâche du *Naturaliste* au Port Jackson (26 avril–18 mai 1802): le témoignage du capitaine Hamelin', *Australian Journal of French Studies*, 41:2 (2004), pp. 126–51, on p. 129.

25. Hamelin, *Journal*, vol. 2, ANF, SM, 5JJ42, entry dated 28–29 floréal an X [18–19 May 1802].

26. 'Lettres, journaux et papers', ANF, SM, 5JJ24. In this list, the document is entitled: 'Notes sur les établissements des Anglais dans la mer du sud, par Hamelin'.

27. Hunt and Carter, *Terre Napoléon*, p. 23.

28. Philip Gidley King to Nicolas Baudin, Sydney, dated 20 June 1802, ANF, SM, 5JJ53.

29. Philip Gidley King to Nicolas Baudin, Sydney, 21 June 1802, ANF, SM, 5JJ53. This letter has been reproduced in Bladen (ed.), *HRNSW*, vol. 4, p. 949, but therein has been presumed incorrectly to be a reply to Baudin's letter of 23 June 1802.

30. Saint-Cricq, *Journal*, ANF, SM, 5JJ48, entry dated 5 floréal an X [25 April 1802].

31. *Journal nautique tenu pendant la campagne de découvertes commandée par le Capitaine de vaisseau Nicolas Baudin, à bord de la corvette le Géographe, par Monsieur Ronsard, officier du génie maritime et lieutenant de vaisseau*, vol. 1, ANF, SM, 5JJ29, entry dated 3 messidor an X [22 June 1802].

32. M. Flinders, 'Journal on the Investigator', January 1801–July 1802, vol. 1, entry dated 22 June 1802, p. 499, Mitchell Library (hereafter ML), Sydney, safe 1/24.

33. Nicolas Baudin to Philip Gidley King, Port Jackson , 22 June 1802, in Bladen (ed.), *HRNSW*, vol. 4, pp. 947–8.

34. Philip Gidley King to Nicolas Baudin, Port Jackson, 23 June 1802, in Bladen (ed.), *HRNSW*, vol. 4, pp. 948–9, on p. 948.

35. Flinders, 'Journal on the Investigator', vol. 1, entry dated 23 June 1802, p. 499.

36. 'Aucune embarcation destinée pour la pêche ne passera à l'ouest de la corvette mais seulement à l'est, on pourra pêcher dans toutes les anses qui sont au nord et au sud, dans ce dernier cas il est expressément défendu de mettre à terre sans une permission écrite et signée du Gouverneur, laquelle désignera les lieux où il sera permis d'aborder'. Ronsard, *Journal*, vol. 1, ANF, SM, 5JJ29, entry dated 3 messidor an X [22 June 1802].

37. 'Il sera désigné un endroit pour laver le linge ainsi que pour faire du bois à brûler' and 'Tous ceux qui par leur conduite ou curiosité indiscrète se mettront dans le cas d'être réprimandé seront renvoyés à bord et consignés'. Ronsard, *Journal*, vol. 1, ANF, SM, 5JJ29, entry dated 3 messidor an X [22 June 1802].

38. E. Scott, *Terre Napoleon: A History of French Explorations and Projects in Australia* (London: Taylor and Francis, 1910), p. 204.

39. Minister of Marine and the Colonies to Nicolas Baudin, Paris, 2 vendémiaire, an IX [29 September, 1800], in Baudin, *The Journal*, pp. 8–9.

40. Nicolas Baudin to Philip Gidley King, Port Jackson, 23 June 1802, in Bladen (ed.), *HRNSW*, vol. 4, p. 949.

41. Ronsard, *Journal*, vol. 1, ANF, SM, 5JJ29, entry dated 4 messidor, an X [23 June 1802].

42. *Journal de Hyacinthe de Bougainville*, ANF, 155 AP6, entry dated 6 messidor an X [25 June 1802]; Ronsard, *Journal*, vol. 1, ANF, SM, 5JJ29, entry dated 6 messidor an X [25 June 1802] and *Table de loch* of the *Géographe*, 8 prairéal to 13 méssidor, an X, ANF, SM, 5JJ25, entry made by H. Freycinet and dated 6 messidor an X [25 June 1802].

43. *Table de loch* of the *Géographe*, 13 messidor to 4 fructidor, an X, entry made by Ronsard and dated 13 messidor [2 July 1802].

44. Bougainville, *Journal*, ANF, 155 AP6, entry dated 6 messidor an X [25 June 1802; Ronsard, *Journal*, vol. 1, ANF, SM, 5JJ29, entry dated 6 messidor an X [25 June 1802] and *Table de loch* of the *Géographe*, 8 prairéal to 13 méssidor, an X, ANF, SM, 5JJ25, entry made by H. Freycinet and dated 6 messidor an X [25 June 1802].

45. *Table de loch* of the *Géographe*, 8 prairial to 13 messidor an X, ANF, SM, 5JJ25, entry made by L. C. G. Bonnefoy and dated 7 messidor an X [26 June 1802].

46. *Table de loch* of the *Géographe*, 8 prairial to 13 messidor an X , ANF, SM, 5JJ25, entry made by Ronsard and dated 9 messidor an X [28 June 1802]; *Table de loch* of the *Géographe*, 13 messidor to 4 fructidor an X, ANF, SM, 5JJ25, entry made by Ronsard and dated 24 thermidor an X [12 August 1802] and *Table de loch* of the *Géographe*, 4 fructidor to X au 24 vendémiaire an XI, ANF, SM, 5JJ25, entry made by H. Freycinet and dated 21 fructidor an X [8 September 1802].

47. 'avec une vitesse stupéfiante dès qu'ils peuvent reprendre une nourriture riche en produits frais'. G. Rigondet, *François Péron 1775–1810 et l'expédition du commandant Nicolas Baudin: les Français à la découverte de l'Australie* (Charroux: Éditions des Cahiers Bourbonnais, 2002), pp. 223–34.

48. *Table de loch* of the *Géographe*, 8 prairéal to 13 méssidor an X, ANF, SM, 5JJ25, entry made by Ransonnet and dated 3 messidor an X [22 June 1802]. Matthew Flinders also commented that the British and French doctors cooperated to treat the patients from the expedition, see M. Flinders, 'Handwritten extract from the Madras Gazette of March 15 1806', The Flinders Papers: letters and documents about the explorer Matthew Flinders (1774–1814), National Maritime Museum, Greenwich, FLI07.

49. *Table de loch* of the *Géographe*, 13 messidor to 4 fructidor, an X, ANF, SM, 5JJ25, entry made by H. Freycinet and dated 13 messidor an X [2 July 1802].

50. *Table de loch* of the *Géographe*, 13 messidor au 4 fructidor, an X, ANF, SM, 5JJ25, entry made by C. Baudin and dated 14 messidor, an X [3 July 1802].

51. *Table de loch* of the *Géographe*, 7 prairial au 13 messidor, an X, ANF, SM, 5JJ25, entry made by Bonnefoy and dated 12 messidor an X [1 July 1802].

52. Bougainville, *Journal*, entry dated 8 messidor an X [27 June 1802].

53. See H.-J. Taillefer, *Sur la Dysenterie observée dans les pays chauds* (Paris: Didot Jeune, 1807), p. 22, cited in J. Southwood and D. Simpson, 'Baudin's Doctors: French Medical Scientists in Australian Waters, 1801–1803', *Australian Journal of French Studies*, 41:2 (2004), pp. 152–64, on p. 158.

54. Baudin, 'Compte général de mes dépenses relatives aux bâtiments de la République, le *Géographe*, le *Naturaliste* et le *Casuarina* pendant la relâche au Port Jackson, Nouvelle-Hollande', ANF SM 5JJ53.

55. Sodium sulfate, which was used as a laxative.

56. Lint was used as a medical dressing.

57. This product seems to have been used as an enema.

58. T. Jamison, 'Supplied the French Corvette *Géographe* Commanded by Commodore Baudin on discovery with the following Medicines from His Majesty's Stores', compiled in Sydney and dated November 2 1802, ANF, SM, 5JJ24 and Baudin, 'Compte général de mes dépenses', ANF, SM, 5JJ53.

59. This is usually spelt 'cinchona' and refers to the bark from any of several trees or shrubs of the genus *Cinchona*. It is a source of quinine and is used to treat malaria.

60. Matthew Flinders to Nicolas Baudin, Port Jackson, undated, ANF, SM, 5JJ53.

61. These reports are each entitled 'Situation des hommes actuellement à l'hôpital de Sidney' and are dated daily from 12 messidor to 26 messidor an X [1 July–15 July 1802] and signed Mr Taillefer, ANF, SM, 5JJ24.

62. 'les autres me paraissent mener une vie qui ne les fera pas jouir longtemps de la santé que leur ont d'abord rendue une nourriture meilleure', 'sans ces précautions le libertinage et la gourmandise convertiront les plus légères indispositions en des maladies interminables'. François Lharidon to Nicolas Baudin, Sydney, 15 messidor an X, [4 July 1802], ANF, SM, 5JJ24.

63. *Table de loch* of the *Géographe*, 4 fructidor an X to 24 vendémiaire an XI, entry made by Ronsard and dated 6 fructidor an X [24 August 1802].

64. Baudin paid a total of £167.17*s* for the medical care provided from 22 June until 24 August. See Baudin, 'Compte général de mes dépenses', ANF, SM, 5JJ53. Concerning the establishment of the medical tent, see Bougainville, *Journal*, entry dated 6 Messidor an X [2 August 1802], p. 21.

65. P.-B. Milius, *Récit du voyage aux terres australes de Pierre-Bernard Milius, second sur le* Naturaliste *dans l'expédition Baudin (1800–1804)*, ed. J. Bonnemains and P. Haugel (Le Havre: Société havraise d'études diverses, 2000), p. 43.

66. 'elles me firent regretter le départ du Naturaliste qui ne pouvait nous rencontrer que par un hasard bien peu réfléchi eu égard à la saison où nous nous trouvions alors'. Nicolas Baudin to Antoine-Laurent de Jussieu, Port Jackson, 20 brumaire an X [11 November 1802], MNHN, ms 2082, part 8.

67. Nicolas Baudin to Antoine Laurent de Jussieu, Port Jackson, 20 brumaire an X [11 November 1802], MNHN, ms 2082, part 5.

68. See, for example, A. Brown, *Ill-Starred Captains: Flinders and Baudin* (Fremantle: Fremantle Press, 2004), p. 254 and Fornasiero, Monteath and West-Sooby, *Encountering Terra Australis*, p. 203.

69. 'il éviterait aussi de rentrer dans une dépendance qui blessait son orgueil'. Saint-Cricq, *Journal*, entry dated 27 floréal an X [17 May 1802].

70. For a discussion of the relationship between Baudin and Hamelin, see Fornasiero, Monteath and West-Sooby, *Encountering Terra Australis*, p. 203.

71. Fornasiero, Monteath and West-Sooby, *Encountering Terra Australis*, p. 203.

72. For an overall illustration of this line of command, as well as the variety of tasks and other activities undertaken by the sailors, see the entries recorded during the Port Jackson sojourn in the logbook aboard the *Géographe*: *Table de loch*, 8 prairial to 13 messidor an X [28 May–2 July 1802], 13 messidor to 4 fructidor an X [2 July–22 August 1802], 4 fructidor an X to 24 vendémiaire an XI [22 August–16 October 1802] and 25 vendémiaire to 7 frimaire, an XI [17 October–28 November 1802].

73. The entries in the logbook of the *Géographe* regularly note the departure of sailors to collect bread and provisions. On 3 July 1802 midshipman Charles Baudin wrote that, due

to the rations ordered by the governor, the French could only have fresh bread five times in each period of ten days. See *Table de loch* of the *Géographe*, 13 messidor to 4 fructidor an X, entry made by C. Baudin and dated 14 messidor [3 July 1802].

74. Couture's daily journal entries, covering the first month of the Port Jackson sojourn and the *Géographe*'s logbooks, illustrate this cooperation most clearly.

75. Il fut impossible de rien faire dès que les matelots eurent de l'argent ... Le soir tout l'équipage était ivre'. Ronsard, *Journal*, vol. 1, ANF, SM, 5JJ29, entry dated 26 thermidor [14 August 1802]. See also *Table de loch* of the *Géographe*, du 13 messidor au 4 fructidor, an X, entry made by Bonnefoy and dated 26 thermidor [14 August 1802].

76. For examples of the sailors' misconduct, punishment and Baudin's management of discipline aboard the Géographe, see *Table de loch* of the *Géographe*, 8 prairial to 13 messidor, an X [28 May–2 July 1802], 13 messidor to 4 fructidor an X [2 July–22 August 1802], 4 fructidor an X to 24 vendémiaire an XI [22 August–16 October 1802] and 25 vendémiaire to 7 frimaire an XI [17 October–28 November 1802]. For examples of punishments issued aboard the *Naturaliste,* see *Journal tenu à bord de la Corvette le* Naturaliste *par J.V.ᵉ Couture Aspirant de 1ᵉʳᵉ Classe*, ANF, SM, 5JJ57, entry dated 4–5 floréal an X [24–5 April 1802], and *Journal [Nautique] du Naturaliste Giraud aspirans de la marine pendant les années 9. 10 & 11*, ANF, SM, 5JJ57, entry dated 4–5 floréal an X [24–5 April 1802].

77. *Table de loch* of the *Géographe*, 13 messidor to 4 fructidor an X, entry made by H. Freycinet and dated 26 messidor an X [14 July 1802].

78. *Table de loch* of the *Géographe*, 13 messidor to 4 fructidor, an X, entry made by Ronsard and dated 27 messidor an X [15 July 1802] and Breton, *Journal*, ANF, SM, 5JJ57, entry dated 26 messidor an X [14 August 1802].

79. *Table de loch* of the *Géographe*, 13 messidor to 4 fructidor, an X, entry made by Ronsard and dated 27 messidor an X [15 July 1802].

80. Nicolas Baudin to Philip Gidley King, aboard the *Géographe* at Port Jackson, 17 July 1802, in Bladen (ed.), *HRNSW,* vol. 4, p. 954 and Philip Gidley King to Nicolas Baudin, Sydney, 17 July 1802, in Bladen (ed.), *HRNSW,* vol. 4, p. 954.

81. *Table de loch* of the *Géographe*, 13 messidor to 4 fructidor, an X, entry made by Ransonnet and dated 1 thermidor an X [20 July 1802]; Ronsard, *Journal*, vol. 1, ANF, SM, 5JJ29, entry dated 1 thermidor an X [20 July 1802]; Breton, *Journal*, entries dated 1 and 2 thermidor an X [20 and 21 July 1802].

82. *Table de loch* of the Géographe, 8 prairial to13 messidor an X [28 May–2 July 1802], 13 messidor to 4 fructidor an X [2 July–22 August 1802], 4 fructidor an X to 24 vendémiaire an XI [22 August–16 October 1802] and 25 vendémiaire to 7 frimaire an XI [17 October–28 November 1802].

83. Bougainville, *Journal*, entry dated 24 messidor an X [13 July 1802].

84. That is, six months since their last port of call: Timor.

85. N. Baudin, 'Note de l'argent que j'ai donné aux etats majors et naturalistes faisant partie de l'expédition / relâche au Port Jackson', ANF, SM, 5JJ24; N. Baudin, 'Etat détaillé des sommes donnés aux officiers pendant le cours de la campagne', ANF, SM, BB4997; N. Baudin, 'Recensement de l'argent donné pendant la relâche au Port Jackson' and N. Baudin, 'Note des billets que j'ai donnés aux naturalistes et autres à payer par Mr Simeon Lord', ANF, SM, 5JJ24.

86. John Franklin to his sister Ann, aboard the *Investigator* at Port Jackson, 18 October 1802, in Franklin family – Letters from John Franklin to family members, 1802–3, ML, citation no. C 231.

87. J. Fornasiero and J. West-Sooby, 'A Cordial Encounter? The Meeting of Matthew Flinders and Nicolas Baudin (8–9 April 1802)', in I. Coller, H. Davies and J. Kalmann (eds), *History and Civilization: Papers from the George Rudé Seminar* (2005), vol. 1, pp. 53–61, on p. 60.

88. Péron, *Voyage of Discovery*, vol. 1, p. 295.

89. J. H. Tuckey, 'A Sketch of the Present State of the Colony of New South Wales, Commercial and Civil, by J. H. Tuckey, 1st Lieutenant of HMS Calcutta', James Hingston Tuckey – Papers, 1804, ML, microfilm – CY 1249, frames 58–122.

90. Baudin, 'Compte général de mes dépenses', ANF, SM, 5JJ53.

91. Baudin, 'Note de l'argent que j'ai donne aux etats majors et naturalistes faisant partie de l'expédition / relâche au Port Jackson', ANF, SM, 5JJ24.

92. Starbuck, 'The Colonial Field'.

93. N. Baudin, 'Commodore Baudin to J. Underwood, 1802', ANF, SM, 5JJ53.

94. Lesueur also produced several sketches of the townships of Parramatta and Brickfield and of Aborigines in natural settings; however, scholars believe that these are copies of, or were at least heavily influenced by, similar illustrations published in D. Collins, *An Account of the British Colony in New South Wales, with Remarks on the Dispositions, Customs, Manners, etc. of the Native Inhabitants of that Country*, vols 1 and 2 (London: A. H. & A. W. Reed, 1802). See R. Jones, 'Images of Natural Man', in J. Bonnemains, E. Forsyth and B. Smith (eds), *Baudin in Australian Waters: The Artwork of the French Voyage of Discovery to the Southern Lands 1800–1804* (Oxford and Melbourne: Oxford University Press, 1988), pp. 35–64, on p. 58.

95. C. P. Boullanger, *Registre tenu par le C^en Boullanger, l'un des ingénieurs géographe, emploié dans l'expédition du Capitaine de vaisseau Baudin aux Terres australes*, ANF, SM, 5JJ44 and P. Faure, *Journal de M. Faure, ingénieur géographe, 1801–1803*, ANF, SM, 5JJ56.

96. 'Plan de la ville de Sydney (Capitale des colonies Anglaises aux Terres australes), levé par Mr. Lesueur & assujetti aux relèvemens de Mr. Boullanger (Novembre, 1802)', plate 30 in L. Freycinet, *Voyage de découvertes aux terres australes exécuté par ordre de sa majesté l'empereur et roi, sur les corvettes le Geographe, le Naturaliste; et la goélette le Casuarina, pendant les années 1800, 1801, 1802, 1803, et 1804*, vol. 3, *Navigation et géographie, Atlas* (Paris: Imprimerie Royale, 1812).

97. P.-F. Bernier, *Observations Astronomique faites pendant l'expédition de Découvertes commandée par le Capitaine Baudin par Pierre François Bernier, Astronome sur la Corvette le Naturaliste*, ANF, SM, 5JJ47.

98. In October 1802, King reported to Joseph Banks that Baudin was planning 'to pass thro' Bass's Straits, and in some part of them he means to land for the purpose of observing the transit of Mercury'. See Philip Gidley King to Joseph Banks, Sydney, 2 October 1802, in Bladen (ed.), *HRNSW*, vol. 4, p. 844.

99. 'Le Commandant m'a donné l'ordre de rester à terre nuit et jour pour l'aider dans ses occupations, en consequence j'ai fait descendre mon lit à terre et je viendrai à bord chaque jour'. Hamelin, *Journal*, vol. 2, ANF, SM, 5JJ42, entry dated 3 thermidor an X [22 July 1802].

100. Hamelin, *Journal*, vol. 2, ANF, SM, 5JJ42, entry dated 27–28 germinal an X [17–18 April 1802] and Emmanuel Hamelin to Nicolas Baudin, Port Jackson, 27 floréal an X in Hamelin, *Journal*, vol. 2, ANF, SM, 5JJ42, entry dated 27 floréal an X [17 April 1802].

101. Baudin, 'Recensement de l'argent donné pendant la relâche au Port Jackson'.

102. Hamelin, *Journal*, vol. 2, ANF, SM, 5JJ42, entry dated 14 thermidor an X [2 August 1802].

103. Ibid., entries dated 1–10 and 20–30 fructidor an X [19–28 August and 7–17 September 1802].
104. Ibid., entry dated 1–5ème complémentaire an X [18–22 September 1802].
105. Ibid., entry dated 19 vendémiaire an XI [11 October 1802].
106. Ibid., entry dated 5 vendémiaire an XI [27 September 1802].
107. Captain William Kent was a naval officer and, in 1802, had been made a magistrate. He was a nephew of John Hunter, the previous governor of New South Wales.
108. Hamelin, *Journal*, vol. 2, ANF, SM, 5JJ42, entry dated 23 brumaire an XI [14 November 1802].
109. Nicolas Baudin to Antoine-Laurent de Jussieu, Port Jackson, 20 brumaire an XI, MNHN, ms 2082, part 8.
110. Henry Underwood and James Kable to Nicolas Baudin, 7 July 1802, ANF, SM, 5JJ53.
111. Philip Gidley King to Nicolas Baudin, Sydney, 11 July 1802, in Bladen (ed.), *HRNSW*, vol. 4, p. 953.
112. 'des terres qu'on dit exister au Nord des Isles Hunter'. Nicolas Baudin to Antoine-Laurent de Jussieu, Port Jackson, 20 brumaire an XI, MNHN, ms 2082, part 8.
113. Nicolas Baudin to Antoine-Laurent de Jussieu, Port Jackson, 20 brumaire an XI, MNHN, ms 2082, part 8.
114. 'Je ... vais faire tous mes efforts pour compléter de nouveau une collection aussi nombreuse que celle que vous allez recevoir par le *Naturaliste*'. Nicolas Baudin to Antoine-Laurent de Jussieu, Port Jackson, 20 brumaire an XI, MNHN, ms 2082, part 5.
115. 'la Nouvelle expédition'. *Journal Tenu Par le lieut^t de V^au L^is Freycinet commandant la goëlette Le* Casuarina. *Mois de Vendémiaire et de Brumaire an XI*, entry undated, ANF, SM, 5JJ49.
116. L. Freycinet, *Journal*, entries 1–8, 22–26 vendémiaire, 3–11, 23–25 brumaire an XI [23–30 September, 14–18 October, 25 October, 14–16 November (1802)].
117. S. Skinner, 'Commodore Baudin, Account of Samuel Skinner', ANF, SM, 5JJ53.
118. N. Baudin, 'Copie des lettres d'échange tirés au Port Jackson o/c de l'Expédition', ANF, SM, BB4997.
119. Anon., 'Commodore Rum Account', ANF, SM, 5JJ53.
120. 'vous savez que, bien loin de donner du rum à sa valeur dans le pays, je l'ai placé sur le prix de 10 schelin, afin que les personnes qui m'ont procuré des objets d'histoire naturelle ou des remplacements en vivres y trouvassent un bénéfice qui put les engager à nous bien servir'. Nicolas Baudin to Philip Gidley King, aboard the *Géographe* at Port Jackson, 4 October 1802, in Bladen (ed.), *HRNSW*, vol. 4, pp. 976–7.
121. 'je crains que tant d'ouvrages ne prennent beaucoup plus de temps que ne le permettront les provisions que nous avons faites ici, car les observations géographiques exigent beaucoup de temps; et toutes reconnaissances faites trop promptement seront superficielles, imparfaites et remplies d'erreurs'. Nicolas Baudin to Antoine-Laurent de Jussieu, Port Jackson , 20 brumaire an XI, MNHN, ms 2082, part 5.
122. 'la colonie du Port Jackson doit fixer l'attention du Gouvernement et même des autres puissances de l'Europe surtout de l'Espagne'. Nicolas Baudin to the Minister of Marine and the Colonies, Kupang, 9 Prairial an XI [29 May 1803], ANF, SM, BB4995.
123. Nicolas Baudin to Antoine-Laurent de Jussieu, Port Jackson, 20 brumaire an XI [11 November 1802], MNHN, ms 2082, part 5.
124. See the comprehensive collection of expense accounts and correspondence relating to the Port Jackson sojourn held at the ANF, SM, 5JJ53.

125. 'Toute la campagne dans le moment où je vous écris est en fleurs dont rien n'égale la beauté et le coup d'œil; je ne connais pour la variété que le Cap de Bonne Espérance qui puisse lui être comparé'. Nicolas Baudin to Antoine-Laurent de Jussieu, Port Jackson, 20 brumaire an XI, MNHN, ms 2082, part 5.

126. 'La plupart et c'est le plus grand nombre se sont retirés très avant dans l'intérieur du pays où ils continuent de vivre à leur manière; d'autres se promènent habituellement dans la ville et dans la campagne'. Nicolas Baudin to the Minister of Marine and the Colonies, Port Jackson, 20 brumaire an XI [11 November 1802], ANF, SM, BB4995.

3 Disciplining Passions: French Naval-Voyagers at Anchor

1. G. Dening, *Mr Bligh's Bad Language: Passion, Power and Theatre on the Bounty* (Cambridge: Cambridge University Press, 1992), pp. 82 and 121.

2. Dening, *Performances*, pp. 108–9.

3. Cormack, *Revolution and Political Conflict in the French Navy*, pp. 297–301.

4. Ibid., p. 2.

5. See M. A. Conley, *From Jack Tar to Union Jack: Representing Naval Manhood in the British Empire, 1870–1918* (Manchester: Manchester University Press, 2009), pp. 4–6.

6. *Table de loch* of the *Géographe*, 8 prairial to 13 messidor, An X, ANF, SM, 5JJ25, entry made by Ronsard and dated 3 messidor [22 June 1802].

7. J. Gascoigne, *Captain Cook: Voyager between Worlds* (London: Continuum Books, 2007), p. 60.

8. Minister of Marine and the Colonies to Nicolas Baudin, Paris, 7 vendémiaire an IX [29 September 1800], in Baudin, *The Journal*, p. 7. Also in *Table de loch* of the *Géographe*, ANF, SM, 5JJ25, inserted at the opening for 16–17 ventôse an IX [7 March 1801].

9. There exists another version of these instructions which, though dated the same, does not include this clause. It has therefore been suggested that Baudin forged the above passage. However, while that is an interesting theory, no evidence has yet come to light to support it. The addition of the clause quoted here is only one difference among many and, overall, this version of the regulations that Baudin used at Port Jackson was not controversial. See the letter from the Minister of Marine and the Colonies, written in Paris and dated 7 vendémiaire an IX [29 September 1800], ANF, SM, BB49995, reproduced in Baudin, *Mon voyage*, pp. 98–100, and E. Duyker, *François Péron: An Impetuous Life* (Melbourne: Miegunyah Press, 2006), p. 67.

10. *Journal nautique tenu pendant la campagne de découvertes commandée par le Capitaine de vaisseau Nicolas Baudin*, ANF, SM, 5JJ29, entry dated 3 messidor, an X [22 June 1802].

11. N. A. M. Rodger, *The Wooden World: An Anatomy of the Georgian Navy* (London: Collins, 1986), p. 207.

12. Conley, *From Jack Tar to Union Jack*, p. 2.

13. C. E. Forth (ed.), *Masculinity in the Modern West: Gender, Civilization and the Body* (Basingstoke: Palgrave Macmillan, 2008), p. 46.

14. *Tables de loch* of the *Géographe*, 8 prairial to 13 messidor an X [28 May–2 July 1802], 13 messidor to 4 fructidor an X [2 July–22 August 1802], 4 fructidor an X to 24 vendémiaire an XI [22 August–16 October 1802] and 25 vendémiaire to 7 frimaire an XI [17 October–28 November 1802], ANF, SM, 5JJ25.

15. *Table de loch* of the *Géographe*, 25 vendémiaire to 7 frimaire, an X, ANF, SM, 5JJ25, entry made by Ransonnet and dated 30 vendémiaire [22 October 1802].

16. *Table de loch* of the *Géographe*, 4 fructidor an X to 24 vendémiaire an XI, ANF, SM, 5JJ25, entry made by Bonnefoy and dated 8 fructidor 922 August 1802]; *Table de loch* of the *Géographe*, 25 vendémiaire to 7 frimaire an XI, ANF, SM, 5JJ25, entry made by Ransonnet and dated 30 vendémiaire [22 October 1802].
17. Dening, *Mr Bligh's Bad Language*, pp. 71–2.
18. *Journal de navigation Du L.*^{t} *de V.*^{au} *H*^{ry} *Freycinet, empbarqué sur la Corvette de La républiquef.*^{se} *Le géographe. An 11 de la R.*^{que} *f.*^{se}, ANF, SM, 5JJ34, entry made 12 brumaire an XI [3 November 1802].
19. Cormack, *Revolution and Political Conflict in the French Navy*, pp. 45–7.
20. Baudin, 'Compte générale de mes dépenses', ANF, SM, BB4997.
21. *Table de loch* of the *Géographe*, 4 fructidor an X to 24 vendémiaire an XI, ANF, SM, 5JJ25, entry made by Bonnefoy and dated 25 fructidor an X [12 September 1802]. On Baudin's comments concerning the importance of relaxation time for sailors, see Baudin, *The Journal*, p. 123.
22. *Table de loch* of the *Géographe*, 13 messidor to 4 fructidor an X, ANF, SM, 5JJ25, entry dated 17 messidor an X [6 July 1802].
23. Dening, *Mr Bligh's Bad Language*, p. 23.
24. *Table de loch* of the *Géographe*, 12 messidor to 4 fructidor an X, ANF, SM, 5JJ25, entries made by Henri Freycinet and dated 19 messidor an X [8 July 1902] and by Bonnefoy, dated 2 fructidor an X [20 August 1802]; *Table de loch* of the *Géographe*, 4 fructidor an X to 24 vendémiaire an XI, ANF, SM, 5JJ25, entry made by Ransonnet and dated 22 fructidor an X [9 September 1802] by Bonnefoy, dated 25 fructidor an X [12 September 1802].
25. Gascoigne, *Captain Cook*, p. 59.
26. Emmanuel Hamelin to Philip Gidley King, aboard the *Naturaliste*, Neutral Harbour and dated 10 floréal an X [13 May 1802], in Bladen (ed.), *HRNSW*, vol. 4, p. 946.
27. *Table de loch* of the *Géographe*, 4 fructidor an X to 24 vendémiaire an XI, ANF, SM, 5JJ25, entry made by H. Freycinet and dated 21 fructidor an X [8 September 1802].
28. Ronsard, *Journal*, vol. 1, ANF, SM, 5JJ29, entry dated 22 fructidor an X [9 September 1802].
29. For a broader discussion on this matter see Cormack, *Revolution and Political Conflict in the French Navy*.
30. Horner, *The French Reconnaissance*, pp. 255–6.
31. Ibid., p. 255–7.
32. G. Monge, *Compte Rendu à la Convention Nationale par le Ministre de la Marine, de l'état de situation de la Marine de la République, le 23 septembre de l'an premier; imprimé et envoyé aux 83 Départements et à l'Armée, par ordre de la Convention Nationale* (Paris: Imprimerie, 1792), p. 2. Quoted in Cormack, *Revolution and Political Conflict in the French Navy*, p. 145.
33. D. De la Roquette, 'Notice historiques sur MM. Henri et Louis Freycinet', *Bulletin de la Société de Géographie*, 20:2 (1843), pp. 501–37, on p. 502.
34. 'Rapport et projet de décret sur le mode d'épurement de la marine civile et militarie, présenté, au nom du Comité de marine, à la Convention nationale, par Topsent, député du Département de l'Eure'; Archives Parlementaires de 1787 à 1860 (Paris: Librairie Administrative Paul Dupont, 1910), p. 185.
35. J. Saint-André, *Rapport des représans du peuple envoyés a Brest et auprès de l'armee navale; par Jeanbon Saint-André* (Paris: Imprimerie Nationale, [n.d.]), p. 10. Quoted in Cormack, *Revolution and Political Conflict in the French Navy*, p. 270.

36. 'Commission de la Marine et des Colonies. Renseignements exigés par l'arrêté du Comité de salut public du 30 germinal pour le nomination de tout emploi au service de la Marine', dated 25 ventôse an III [15 March 1795], Service Historique de la la Marine, Vincennes, Série CC7, dossier personel: Ronsard, François-Michel.

37. 'Rapport sur la marine de la république dans la Méditerranée par Barère, au nom du Comité de salut public, dans la séance du 14 Nivôse an II', AP, vol. 82, p. 615, quoted in Cormack, *Revolution and Political Conflict in the French Navy*, p. 260.

38. Ronsard to the Minister of Marine and the Colonies, dated 5 June 1824, SHM, CC7 Ronsard.

39. C. Baudin, *Souvenirs de jeunesse de l'amiral Baudin*, SHM, CC7a125, ms 116, p. 86.

40. Horner, *The French Reconnaissance*, p. 256.

41. 'ce paysan, votre caporal Péron ... un caporal comme lui vaut 10,000 officiers comme vous ... c'est un cochon ... C'est vous qui êtes un cochon ... j'espère Monsieur que vous voudrez bien rendre au Commandant compte des motifs pour lesquels vous m'avez puni'. This description of the argument is based on Ronsard's journal entry, see Ronsard, *Journal*, vol. 1, ANF, SM, 5JJ29, entry made 4 vendémiaire an X [26 September 1802]. He wrote a more concise version in the logbook, see *Table de loch* of the *Géographe*, 4 fructidor an X to 24 vendémiaire an XI, ANF, SM, 5JJ25, entry made by Ronsard and dated 3 vendémiaire an XI [25 September 1802].

42. 'Comme officier de garde ... votre devoir était, et vous ne l'ignorez pas, d'imposer silence à l'un et à l'autre ou au moins de les inviter à aller discuter ailleurs que dans la grande chambre une question absolument étrangère au service du bâtiment et de pure opinion'. Nicolas Baudin to François-Michel Ronsard, Sydney, 5 vendémiaire an XI [27 September 1802], in Ronsard, *Journal*, vol. 1, ANF, SM, 5JJ29, entry made 5 vendémiaire an XI [27 September 1802].

43. B. Rosenwein, *Emotional Communities in the Early Middle Ages* (Ithaca, NY: Cornell University Press, 2006).

44. G. W. Rusden, *History of Australia*, vol. 1 (Cambridge: Cambridge University Press, 1883), pp. 301–2; M.-S. Rivière, *A Woman of Courage: The Journal of Rose de Freycinet on her Voyage around the World, 1817–1820* (Canberra: National Library of Australia, 1996), p. 26; Duyker, *François Péron*, p. 137.

45. In the first edition of the voyage narrative, Péron made the following statement in relation to his meeting with a French settler in the colony: 'Avec quelle douce satisfaction je m'empressai de raconter à cet intéressant compatriote toute cette suite de prodiges par lesquels un grand homme étoit enfin parvenu à rendre le bonheur et la paix à notre patrie commune!' (With what sweet satisfaction I hastened to relate to this interesting compatriot the series of wonders by which a great man had finally succeeded in returning happiness and peace to our common homeland!) The attribution of France's improvement to 'un grande homme' is not included in the second edition of this narrative. See F. Péron, *Voyage de découvertes aux terres australes exécuté par ordre de Sa Majesté l'Empereur et Roi, sur les corvettes le* Géographe, *le* Naturaliste; *et la goélette le* Casuarina, *pendant les années 1800, 1801, 1802, 1803, et 1804*, vol. 1, *Historique* (Paris: Imprimerie Imperiale, 1807) vol. 1, p. 432. Expressing similar patriotic zeal, on the 1 vendémiaire an XI, the anniversary of the Republique, Louis Freycinet wrote in his journal: 'Quelle epoque plus glorieuse et plus chere à un français républicain! Elle me rappelle vivement les obligations que j'ai contracté comme Militaire, envers ma chere patrie. Ah! je ne les oublirai jamais! Quelles que puissent être Les circonstances où je me trouve je ferai toujours mon devoir. Personne ne pourra se vanter d'avoir pousse à bout et ma patience et mon cour-

age'. (What time more glorious and dear to a French Republican! It reminds me sharply of the obligations that I took on, as a member of the military, to my dear homeland. Ah! I will never forget! No matter what the circumstances I find myself in I will carry out my work. No-one will be able to boast of having pushed me to the limit of my patience and my courage.) L. Freycinet, *Journal*, ANF, SM, 5JJ34, entry dated 1 vendémiarie, an XI [23 September 1802].

46. 'Services à la Mer de Jacques Joseph Ransonnet, Lieutenant de Vaisseau, Membre de Légion d'Honneur' and 'Notice', Service Historique de la Defence, Archives Centrale de la Marine, Vincennes (hereafter SHD, Marine), série CC7: Ransonnet, François-Joseph.

47. 'Extrait de Registre des Actes de l'Etat Civil de la Ville de Laon, Département de l'Aince', SHD, Marine, série CC7: Bonnefoy de Montbazin, Antoine Louis-Gilles.

48. N. Baudin, *Journal du voyage aux Antilles de* La Belle Angélique, *1796–1798*, ed. M. Jangoux (Paris: Press de l'Université Paris-Sorbonne, 2009), p. 41.

49. Ronsard to N. Baudin, reproduced in Ronsard, *Journal*, t. I, ANF, SM, 5JJ29, entry dated 20 vendémiaire an XI [12 October 1802].

50. *Table de loch* of the *Géographe*, 13 messidor to 4 fructidor an X, ANF, SM, 5JJ25, entry dated 17 messidor an X [6 July 1802].

51. See B. J. Martin, *Napoleonic Friendship: Military Fraternity, Intimacy and Sexuality in Nineteenth-Century France* (Durham, NH: University of New Hampshire; University Press of New England, 2011).

52. 'Vous ne devrez jamais oublier que s'il est difficile de commander aux hommes, il est néanmoins glorieux de bien les conduire. On y réussit presque toujours quand on se comporte envers eux avec modération, prudence et justice. D'après ces principes, qui ne sont pas ou très rarement mis enpratique, par de jeunes officiers, vous devez donc éviter toute occasion qui pourrait compromettre votre autorité et même votre personne. Celui qui command n'est point excusable quand se portant à des excès que condamnent les lois, il n'écoute que ses passions et punit arbitrairement, ou par caprice, l'individu qui a mérité de l'être'. Letter from Nicolas Baudin to Louis Freycinet written in Sydney and dated 1 vendemiaire XI [23 September 1802], reproduced in *Journal Tenu Par le lieutenant de Vaisseau Louis Freycinet commandant la goëlette* Le Casuarina. *Mois de Vendémiaire et de Brumaire an XI*, entry undated, ANF, SM, 5JJ49.

53. Dening, *Mr Bligh's Bad Language*, p. 145.

54. Horner, *Looking for La Pérouse*, pp. 87 and 102.

55. Ibid., p. 80.

56. Gascoigne, *Captain Cook*, p. 59.

57. W. Reddy, 'Sentimentalism and its Erasure: The Role of Emotions in the Era of the French Revolution', *Joural of Modern History*, 72:1 (March 2000), pp. 109–52 and *The Navigation of Feeling: A Framework for the History of Emotions* (Cambridge: Cambridge University Press, 2001), pp. 173–210 and pp. 216–7.

58. 'des plaisirs passagers à des devoirs réels et obligatoires ... La campagne que j'ai à faire n'est pas celle d'un bâtiment de guerre où le temps des relâches est le moment des plaisirs et des amusements. C'est au contraire celui d'un travail encore plus actif et laborieux que celui de la mer' ... 'Les occasions de faire le mal sont trop multipliées et trop faciles dans le lieu et l'état où se trouve la corvette pour en agir autrement. Nicolas Baudin to officers of the *Géographe*, in *Table de loch* of the *Géographe*, 13 messidor to 4 fructidor an X, ANF, SM, 5JJ25, entry dated 1 thermidor an X [20 July 1802].

4 The French and the British: A Diplomatic Relationship

1. Lyons, *Napoleon Bonaparte*, pp. 204–5.
2. Joseph Banks to Philip Gidley King, Soho Square, 1 January 1801. King family – Correspondence and memoranda, 1775–1806, ML, reference A 1980/2 CY 906, p. 37.
3. For instance see Horner, *The French Reconnaissance*, pp. 240 and 247; Hunt and Carter, *Terre Napoléon*, p. 23 and Duyker, *François Péron*, p. 138.
4. Minister of Marine and the Colonies to Nicolas Baudin, Paris, 7 vendémiaire an IX [29 September 1800], in Baudin, *The Journal*, p. 9.
5. Minister of Marine and the Colonies to Nicolas Baudin, Paris, 7 vendémiaire an IX [29 September 1800]. Baudin, *The Journal*, p. 8.
6. Dening, *Performances*, p. 109.
7. Ibid.
8. See J. Gascoigne, *Science in the Service of Empire: Joseph Banks, the British State and the Uses of Science in the Age of Revolution* (Cambridge: Cambridge University Press, 1998).
9. Joseph Banks to George John Spencer, Soho Square, December 1800, in N. Chambers (ed.), *The Letters of Sir Joseph Banks: A Selection, 1768–1820* (London: Imperial College Press, 2000), pp. 219–21.
10. Joseph Banks to George John Spencer, p. 219.
11. Ibid., p. 219.
12. Ibid., p. 220.
13. Fornasiero and West-Sooby, 'A Cordial Encounter?', p.61.
14. Matthew Flinders to Joseph Banks, Port Jackson, 20 May 1802, in Bladen (ed.), *HRNSW*, vol. 4, pp. 755–7, on p. 755.
15. Banks explains that the publication was of a letter written by the gardener on board the *Géographe*, Anselm Riedlé, in which Riedlé complains of too seldom having had the opportunity to anchor in a harbour or to go ashore. It is important to note that Riedlé died before the Baudin expedition reached the south coast. His letter could only have concerned the first few months of the expedition's exploration of Australia, which were spent on the west coast where Flinders had not yet surveyed. See Joseph Banks to Matthew Flinders, Soho Square, 10 April 1803, Papers of Sir Joseph Banks 1745–1923 [manuscript], series 1: Correspondence and Papers of Sir Joseph Banks, National Library of Australia (hereafter NLA), ms 9/22.
16. Joseph Banks to Philip Gidley King, Soho Square, 1 January 1801. King family – Correspondence and memoranda, 1775–1806, ML, reference A 1980/2 CY 906, p. 37.
17. See G. Blainey, *The Tyranny of Distance: How Distance Shaped Australia's History* (Melbourne: Sun Books, 1966), p. 50.
18. Philip Gidley King to Lord Hobart, Sydney, 7 August 1803, in F. Watson (ed.), *Historical Records of Australia*, series 1, vol. 4 (Sydney: Library Committee of the Commonwealth Parliament, 1915) (hereafter *HRA*), pp. 357–8.
19. In fact, the British did finally invade and take possession of the Ile de France in 1810.
20. Hamelin, *Journal*, vol. 2, ANF, SM, 5 JJ42, entry dated 7–8 floréal an X [27 April 1802].
21. A. Atkinson, *The Europeans in Australia: A History*, vol. 1 (Oxford: Oxford University Press, 1997), pp. 241–7.
22. Ronsard, *Journal*, vol. 1, ANF, SM, 5JJ29, entry dated 3 messidor an X [22 June 1802]. Ronsard was correct, in that Baudin held the rank of post-captain and King, of commander. For definitions of these ranks see I. C. B. Dear and P. Kemp, *The Oxford*

Companion to Ships and the Sea (Oxford: Oxford University Press, 1976), pp. 125 and 440.

23. 'Je trouve qu'on y parle trop en maître et comme on ferait à un capitaine marchand et non à un Commandant d'expédition'. Ronsard, *Journal*, vol. 1, ANF, SM, 5JJ29, entry dated 3 messidor an X [22 June 1802].

24. John Harris to Philip Gidley King, 25 September 1802, in Bladen (ed.), *HRNSW*, vol. 4, pp. 963–4.

25. John Harris to Nicolas Baudin, 25 September 1802, in Bladen (ed.), *HRNSW*, vol. 4, pp. 964–6, on p. 964.

26. 'La lettre que j'ai adressé à M. Harris, et dont je joint ici une copie, vous mettra à même de juger combien j'ai lieu de me plaindre de la conduite légère et peu réfléchie qu'il a tenu dans le rapport qu'il vous a fait; conduite qu'il a occasionné de ma part une lettre de reproche et de réprimande à des officiers qui étoient loin de l'avoir mériter'. Nicolas Baudin to Philip Gidley King, written at Port Jackson and dated 1 vendémiaire an XI [23 September 1802], in Bladen (ed.), *HRNSW*, vol. 4, pp. 957–8, on p. 957.

27. 'par trop de confiance en ce que vous m'avez dit, j'ai adressé une lettre amère et de reproches à tous mes officiers, tandis que par leur réponse, dont la véracité ne peut être contestée, ils se sont scrupuleusement conformés aux lois d'honneur, de loyauté, et de politesses qui sont la base de leur conduite'. Nicolas Baudin to John Harris, Sydney, 2 vendémiaire an XI [23 September 1802], in Bladen (ed.), *HRNSW*, vol. 4, pp. 958–60, on pp. 958–9.

28. Philip Gidley King to Nicolas Baudin, Sydney, 25 September 1802, in Bladen (ed.), *HRNSW*, vol. 4, pp. 962–3, on p. 963.

29. Nicolas Baudin to the officers of the *Géographe*, Sydney, undated, in the *Table de loch* of the *Géographe*, ANF, SM, 5JJ25, entry dated 2 vendémiaire an XI [24 September 1802].

30. The letters were recorded in the *Table de loch* of the *Géographe*, 4 fructidor to X au 24 vendémiaire Year XI, ANF, SM, 5JJ25, entry dated 1 vendémiaire an X [23 September 1802], and Baudin's letter together with Ronsard's response is reproduced in Ronsard's *Journal*, vol. 1, ANF, SM, 5JJ29, entry dated 1 vendémiaire an X [23 September 1802].

31. 'Si vous voulez avoir la complaisance de parcourir les lois d'honneur de la Marine Fran-çaise, lois que nous avons toujours respectés, vous y vairez *à l'article 11, chapitre 17, page 268, que la place d'honneur que doit occuper le pavillion d'une nation étrangère qu'on veut distinguer doit être placé du côté de Stribord à la grande vergue*. La même loi ajoute, *quand on ne sera pas dans le cas de faire cette distinction, cette même place ne sera jamais occu-pée que par un pavillion français*. Jugez donc Monsieur, si après avoir strictement remplis cette formalité, je n'ai pas le droit de me plaindre amèrement de votre procédé, comme de celui de ceux qui vous ont accompagné, ou aux propos indiscrets qu'on a tenu a ce sujet, propos que l'ignorance de nos usages auroient au moins dû suspendre jusqu'à une plus [p. 959] ample information. Vous pourriez me dire pour vous excuser, ainsi que ceux qui se sont plains, que votre façon de pavoiser les bâtiments n'est pas la même; mais dans ce cas j'aurois à vous répondre que ne la connoiscent pas, je ne me serois jamais permis la moindre observation, et que je n'aurois pu m'imaginer que ce fut par mépris ou tout autre raisons ausi peu concéquentes que vous n'eurriez pas placé le pavillion français dans le lieu établis par nos règlements pour le pavillion de toute nation à laqu'elle on doit des égards'. Nicolas Baudin to John Harris, Port Jackson, 23 September 1802, in Bladen (ed.), *HRNSW*, vol. 4, pp. 958–60.

32. 'Si vous voulez avoir la complaisance de parcourir les lois d'honneur de la Marine Fran-çaise, lois que nous avons toujours respectés, vous y vairez *à l'article 11, chapitre 17, page*

268, que la place d'honneur que doit occuper le pavillion d'une nation étrangère qu'on veut distinguer doit être placé du côté de Stribord à la grande vergue. La même loi ajoute, *quand on ne sera pas dans le cas de faire cette distinction, cette même place ne sera jamais occupée que par un pavillion français.* Jugez donc Monsieur, si après avoir strictement remplis cette formalité, je n'ai pas le droit de me plaindre amèrement de votre procédé, comme de celui de ceux qui vous ont accompagné, ou aux propos indiscrets qu'on a tenu a ce sujet, propos que l'ignorance de nos usages auroient au moins dû suspendre jusqu'à une plus [p. 959] ample information. Vous pouriez me dire pour vous excuser, ainsi que ceux qui se sont plains, que votre façon de pavoiser les bâtiments n'est pas la même; mais dans ce cas j'aurois à vous répondre que ne la connoiscent pas, je ne me serois jamais permis la moindre observation, et que je n'aurois pu m'imaginer que ce fut par mépris ou tout autre raisons ausi peu concéquentes que vous n'eurriez pas placé le pavillion français dans le lieu établis par nos règlements pour le pavillion de toute nation à laqu'elle on doit des égards'. Letter from Nicolas Baudin to John Harris, written at Port Jackson and dated 23 September 1802, in Bladen (ed.), *HRNSW*, vol. 4, pp. 975–7.

33. *Journal de Navigation Du L.*^{t} *deV.*^{au} *h*$^{ry.}$ *Freycinet, embarqué sur la Corvette de La republique f.*^{se} *Le géographe. An 11 dela R.*^{que} *f.*^{se}, ANF, SM, 5JJ34, entry dated 16 vendémiaire an XI [8 October 1802].
34. 'Toute affaire qui attaque l'honneur d'un officier est délicate. Vous savez que le soupçon, même sans fondement, est une injure qui se pardonne difficilement, et je ne vous dissimulerai pas que j'ai été obligé d'employer l'autorité pour éviter une scène'. Nicolas Baudin to Philip Gidley King, aboard the *Géographe,* Port Jackson, 4 October 1802, in in Bladen (ed.), *HRNSW,* vol. 4, pp. 975–7.
35. Philip Gidley King to William Paterson, Sydney, 4 October 1802, in Bladen (ed.), *HRNSW*, vol. 4, p. 980. It is worth noting that this affair caused some conflict between King and Paterson, whose relationship had already begun to deteriorate as they battled to bring about changes within the privileged NSW officer corps. See D. S. Macmillan, 'Paterson, William (1755–1820)', in *Australian Dictionary of Biography*, vol. 2 (Carlton: Melbourne University Press, 1967), p. 318.
36. 'chacun de nous lui parut un spadassin'. Ronsard, *Journal*, vol. 1, ANF, SM, 5JJ29, entry dated 16 vendémiaire an XI [8 October 1802].
37. Ronsard, *Journal*, vol. 1, ANF, SM, 5JJ29, entry dated 16 vendémiaire an XI [8 October 1802].
38. Letter from Anthony Fenn Kemp to Nicolas Baudin, in Ronsard, *Journal*, vol. 1, ANF, SM, 5JJ29, entry dated 16 vendémiaire an XI [8 October 1802].
39. 'Il s'avisa de dire qu'il était fâché d'avoir fait des excuses et qu'il eut préféré se battre'. Ronsard, *Journal*, vol. 1, ANF, SM, 5JJ29, entry dated 16 vendémiaire an XI [8 October 1802].
40. Horner, *The French Reconnaissance*, p. 255 and Brown, *Ill-Starred Captains*, p. 267.
41. For a full account of this incident, see Ronsard, *Journal*, vol. 1, ANF, SM, 5JJ29, entry dated 16 vendémiaire an XI [8 October 1802].
42. 'Je suis persuadé que par la crainte de nous avoir pour voisin ils vont faire occuper cette partie de la Terre de Diémen afin de tacher de constate d'une manière authentique leur droit de propriété'. Nicolas Baudin to the Minister of Marine and the Colonies, Port Jackson, 20 brumaire an X [11 November 1802], ANF, SM, BB4995.
43. Nicolas Baudin to Phillip Gidley King, Elephant Bay, King Island, 3 nivose an XI [24 December 1802], in Bladen (ed.), *HRNSW*, vol. 5, pp. 826–30, on p. 826.

44. Baudin to King, Elephant Bay, King Island, 3 nivose an XI [24 December 1802], in Bladen (ed.), *HRNSW*, vol. 5, p. 826.

45. P. Russell, *Savage or Civilized? Manners in Colonial Australia* (Sydney: University of New South Wales Press, 2010), pp. 85–6.

46. M. McKellar (ed.), *Strangers in a Foreign Land: The Journal of Niel Black and Other Voices from the Western District* (Carlton: Miegunyah Press, 2008), p. 70.

47. Philip Gidley King to Joseph Banks, Sydney, 10 November 1802, ANF, SM, 5JJ53.

48. Nicolas Baudin to the Minister of Marine and the Colonies, Kupang Bay, 9 prairéal an 11 [29 May 1803], ANF, BB4995.

49. 'Lettres, journaux et papiers', ANF, SM, 5JJ24 and Hamelin, *Journal*, vol. 2, ANF, SM, 5JJ42, entry dated 28–29 floréal an X [18–19 May 1802].

50. P.-B. Milius, 'Coup d'oeil rapide sur l'établissement des Anglais à la Nouvelle Hollande', Bibliothèque Municipale de Caen, Archive du général Decaen (hereafter BC, AGD) vol. 92, ms 177, fols 74–8 and F. Péron, 'Rapport de François Péron au Général Decaen sur les Colonies anglaise de la Nouvelle Hollande', BC, AGD, vol. 92, fol. 2, available online at http://sydney.edu.au/arts/research/baudin/written_records/other_documents.shtml [accessed 2 January 2012].

51. Fornasiero, Monteath and West-Sooby, *Encountering Terra Australis*, p. 381. Péron also later made a far more lengthy and comprehensive report, however there is no record that that this was seen by any French authorities. See F. Péron, 'Mémoire sur les établissemens anglais à la Nouvelle-Hollande, à la terre de Diémen et dans les archipels du grand Océan Pacifique, au citoyen Fourcroy membre du conseil d'état', CL, MHN Le Havre, dossier 12 and J. Fornasiero and J. West-Sooby, *French Designs on Colonial New South Wales. François Péron's Memoir on the English Settlements in New Holland, Van Diemen's Land and the Archipelagos of the Great Pacific Ocean* (Adelaide: Friends of the State Library of South Australia, 2013).

52. 'Private Instructions from the King to the Sieur de la Pérouse, Captain in the Navy. Commanding the Frigates the *Boussole* and *Astrolabe*. June 26, 1785', in J.-F. de Galaup de La Pérouse, *A Voyage Round the World, Performed in the Years 1785, 1786, 1787 and 1788 by the* Boussole *and the* Astrolabe, *Under the Command of J. F. G. De la Pérouse*, trans. L. A. Millet-Mureau (London: S. Hamilton, 1799), p. 41.

5 Liberty, Equality and 'Civilization': Observations of Colonial Aborigines

1. Nicolas Baudin to Phillip Gidley King, Elephant Bay, King Island, 3 nivose an XI [24 December 1802], in Bladen (ed.), *HRNSW*, vol. 5, p. 826.

2. H. Morphy, 'Encountering Aborigines', in S. Thomas (ed.), *The Encounter 1802: Art of the Flinders and Baudin Voyages* (Adelaide: Art Gallery of South Australia, 2002), pp. 148–63, on p. 154; Fornasiero and West-Sooby, 'Taming the Unknown', pp. 59–80; Sankey, 'The Aborigines of Port Jackson'.

3. The term 'anthropology', though not in common usage, was applied in 1800 in the sense of a comprehensive 'natural science of man'. See. C. Blanckaert, 'L'anthropologie en France, le mot et l'histoire (XIVe–XIXe siècles)', *Bulletins et Mémoires de la Société d'anthropologie de Paris*, new series, 1:3–4 (1989), pp. 20–2.

4. Degérando, 'l'observation des peuples sauvages', p. 131.

5. Ibid., pp. 129–69.

6. A. Malaspina, 'Loose Notes on the English Colony of Port Jackson', in R. J. King (ed.), *The Secret History of the Convict Colony: Alexandro Malaspina's Report on the British Settlement of New South Wales* (Sydney: Allen & Unwin, 1990), pp. 132–50, on pp. 144–9.

7. L. Freycinet, *Voyage autour du monde, entrepris par ordre du roi, sous le ministère et conformément aux instruction de S. Exc. M. le Vicomte du Bouchage, secrétaire d'état au département de la marine, exécuté sur les corvettes de S. M. l'*Uranie *et la* Physicienne, *pendant les années 1817, 1818, 1819 et 1820, Historique*, vol. 2 (Paris: Pillet Aîné, 1839), pp. 893–908.

8. Blanckaert, '1800 – Le moment "naturaliste"'.

9. Staum, *Minerva's Message*.

10. 'Une seule question pourrait même résumer l'enjeu du temps: si le progrès est bien, comme l'affirme le législateur, la loi du genre humain ou le fait d'une élite mieux douée'. Blanckaert, '1800 – Le moment "naturaliste"', p. 119.

11. 'la voix de la raison ne suffit pas'. Blanckaert, '1800 – Le moment "naturaliste"', p. 118.

12. M. de Certeau, D. Julia and J. Revel, *Une politique de la langue: la révolution française et les patois – l'enquête de Grégoire* (Paris: Gallimard, 1975), p. 151.

13. 'l'envers sordide et laide de sa France imaginaire'. Bourguet, 'Race et folklore', pp. 811–12, 815, 817; Staum, *Minerva's Message*, p. 160 and Blanckaert, '1800 – Le moment "naturaliste"', p. 135.

14. Bourguet, 'Race et folklore', p. 812.

15. Harrison, 'Replotting the Ethnographic Romance', p. 40.

16. 'La Société, en cherchant à relever la dignité humaine, cette belle prérogative qui fut si cruellement méconnue, si insolemment outragée, pendant l'affreux régime qui pesa quelque temps sur la France, aura l'avantage de concourir, par la seule influence de ses observations, à l'extinction d'une foule d'abus que ce régime odieux fit naître, et que le gouvernement actuel n'a pu parvenir encore à détruire complètement'. Jauffret, 'Introduction aux mémoires', in *Aux origines de l'anthropologie*, p. 85. Based largely on his family and academic connections as well as the fact that he fleed Paris during the Reign of Terror, Jean-Luc Chappey suggests that Jauffret, a lawyer, journalist and pedagogue of bourgeois heritage as well as founder of the Société des Observateurs, was an advocate of religious morality during the French Revolution. Jauffret may have expressed a similar sentiment as that expressed here in relation to the *ancien régime*, just as well as to the Terror, but his background, added to the fact that he states that this regime lasted only 'quelque temps' and 'introduced' abuses, does suggest that it was in fact the Terror to which this passage refers. See Chappey, *La société des observateurs de l'homme*, pp. 96–8.

17. See for instance Stocking, 'French Anthropology in 1800'; Bourguet, 'Race et folklore', pp. 802–23; Blanckaert, '1800 – Le moment "naturaliste"'; Chappey, *La société des observateurs de l'homme*, pp. 225–380; Harrison, 'Replotting the Ethnographic Romance'.

18. 'laissant de côté toutes ces vaines théories, toutes ces spéculations hasardée'. Jauffret, 'Introduction aux mémoires', in *Aux origines de l'anthropologie*, p. 71.

19. 'Cette direction particulière lui offrira les recherches les plus neuves, les plus importante.' Jauffret, 'Introduction aux mémoires', in *Aux origines de l'anthropologie*, p. 71.

20. F. Péron, 'Observations sur l'anthropologie ou l'histoire naturelle de l'homme', in Copans and Jamin (eds), *Aux origines de l'anthropologie*, pp. 177–85.

21. 'Private Instructions From the King to the Sieur de La Pérouse, Captain in the Navy, Commanding the Frigates *La Boussole* and *Astrolabe*, June 26 1785' ('Part the Fourth: Of the Conduct to be Observed Towards the Natives of the Countries Where the Two

Frigates May Land'), in J.-F. de Galaup de la Pérouse, *A Voyage Round the World*, vol. 1, pp. 50–4.

22. Sankey, 'The Baudin Expedition in Port Jackson, 1802', p. 6. The term 'anthropology', though not in common usage, was applied in 1800 in the sense of a comprehensive 'natural science of man'. See. Blanckaert, 'L'anthropologie en France' and Sankey, 'The Baudin Expedition in Port Jackson, 1802', p. 6.

23. Degérando, 'l'observation des peuples sauvages' and Cuvier, 'Note Instructive', in *Aux origines de l'anthropologie*, pp. 129–69.

24. N. Thomas, 'Introduction', in N. Thomas and D. Losche (ed.), *Double Vision: Art Histories and Colonial Histories in the Pacific* (Cambridge: Cambridge University Press, 1999), pp. 1–16, on p. 5.

25. N. Baudin, 'Des naturels que nous trouvions et de leur conduite envers nous', in *Aux origines de l'anthropologie*, pp. 205–17, on p. 207.

26. Baudin, 'Des naturels', in *Aux origines de l'anthropologie*, pp. 207–17.

27. Péron, *Voyage of Discovery*, vol. 1, pp. 353–4, 359, 364–9; Konishi, *The Aboriginal Male*, pp. 83–5.

28. 'que comme botaniste' and 'son imagination fertile'. Baudin, 'Des naturels', in *Aux origines de l'anthropologie*, pp. 207 and 210.

29. Péron, *Voyage of Discovery*, vol. 1, p. 353.

30. Ibid., pp. 351–85.

31. Nicolas Baudin to Antoine-Laurent de Jussieu, Port Jackson, 20 brumaire an X [11 November 1802], MNHN, ms 2082, part 5.

32. T. Leschenault, 'Account of the Vegetation of New Holland and Van Diemen's Land', in F. Péron [L. Freycinet], *Voyage of Discovery to the Southern Lands*, vol. 3, *Dissertations on Various Subjects*, trans. C. Cornell (Adelaide: The Friends of the State Library of South Australia, 2007), pp. 97–109, on p. 108.

33. F. Péron, 'Conférénce adressée à "Messieurs les Professeurs" décrivant les aborigènes et leur moeurs près de Port Jackson', CL, MHN Le Havre, dossier 09 032.

34. Atkinson, *The Europeans in Australia*, vol. 1, pp. 165–6. G. Karskens, *The Colony: A History of Early Sydney* (Crows Nest: Allen & Unwin, 2009), pp. 478–80.

35. Karskens, *The Colony*, p. 480.

36. Ibid., p. 480.

37. Ibid., pp. 479–80.

38. Ibid., p. 479.

39. Ibid., p. 480.

40. 'leur tendre la main pour s'élever à un état plus heureux! ... le pacte d'une fraternelle alliance! ... Portez-leur nos arts, et non notre corruption, le code de notre morale et non l'exemple de nos vices, nos sciences, et non pas notre scepticisme, les avantages de la civilisation, et non pas ses abus; cachez-leur qu'en ces contrées aussi, quique plus éclairées, les hommes s'entre-déchirent pars des combats, et se dregradent par leurs passions'. Degérando, 'Considérations', p. 132.

41. It is worth noting here that, as mentioned by Jonathan Lamb, Vanessa Smith and Nicolas Thomas, the accumulative effect of closer colonial contacts and incidents of violent resistance to colonization was less romanticized and more derogatory representations of Indigenous peoples. These representations stressed the colonizeds' supposed 'racial' inferiority rather than their social simplicity. See J. Lamb, V. Smith and N. Thomas, *Exploration and Exchange: A South Seas Anthology, 1689–1900* (Chicago, IL: University of Chicago Press, 2000), p. xvii.

42. Bourguet, 'Race et folklore', p. 185.
43. 'la civilization n'ait fait aucun progrès parmi ces peuples depuis plus de quinze ans que les anglais habitent cette isle'. Milius, *Récit du voyage aux terres australes par Pierre-Bernard Milius, second sur le* Naturaliste *dans l'expédition Baudin (1800–1804)*, ed. J. Bonnemains and P. Haugel (Le Havre: Société havraise d'études diverses, 1987), p. 48.
44. Leschenault, 'Account of the Vegetation of New Holland', p. 107.
45. 'aucun désir de changer de condition ... leur penchant naturel est l'indolence' Milius, *Récit du voyage*, p. 48.
46. Milius, *Récit du voyage*, p. 48.
47. 'D'après la répugnance que Banedou a temoigné pour nos usages, il est impossible d'esperer de ramener les sauvages de ce pays à quelques idées de civilization. Ce sont de veritables bêtes brutes qu'il faut laisser vivre à leurs manières'. Milius, *Récit du voyage*, p. 49.
48. Konishi, *The Aboriginal Male*, pp. 127–42.
49. Péron, *Voyage of Discovery*, vol. 1, pp. 351–85.
50. Ibid., pp. 299–300, 303–4 and 305.
51. Ibid., pp. 301, 302, 311, 313 and chapter 20, 'Experiments on the Physical Strength of the Native Peoples of Van Diemen's Land and New Holland the Inhabitants of Timor', pp. 351–85.
52. M. Staum, 'The Paris Geographical Society Constructs the Other, 1821–1850', *Journal of Historical Geography*, 26:2 (2000), pp. 222–38.
53. Péron stated: 'All of New Holland, from Wilson's Promontory in the south to Cape York in the north, appears to be inhabited by a second race of men, differing essentially from those known up until the present day'. Péron, *Voyage of Discovery*, vol. 1, p. 354.
54. Scholars debate just how racialized was Péron's thinking about human diversity. In particular, his claim that the Aboriginal peoples of Tasmania and New Holland were of different 'origins' has incited discussion concerning whether or not Péron believed in polygenism. For a summary of this issue, see J. Fornasiero, 'Deux observateurs de l'homme aux antipodes: Nicolas Baudin et François Péron', in M. Jangoux (ed.), *Portés par l'air du temps: les voyages du Capitaine Baudin*, special number of *Études sur le 18ème siècle*, 38 (2010), pp. 157–70, on pp. 163–5. For a comprehensive study of the evolution from universalist thinking about human nature to the 'science of race', see B. Douglas, 'Climate to Crania: Science and the Racialization of Human Difference, 1750–1880', in B. Douglas and C. Ballard (eds), *Foreign Bodies: Oceania and the Science of Race, 1750–1940* (Canberra: ANU E Press, 2008), pp. 33–96.
55. Péron, *Voyage of Discovery*, vol. 1, pp. 359–60.
56. Ibid., pp. 361–9.
57. Ibid., p. 367.
58. Ibid., pp. 366–7.
59. Ibid., pp. 367–9.
60. Péron, *Voyage de découvertes*, p. 368.
61. L. P. Rivière, 'Un périple en Nouvelle Hollande au début du XIXᵉ siècle', *Comptes-rendus mensuels des séances de l'Académie des Sciences coloniales*, 13 (1953), pp. 571–89, on p. 580 and C.-A. Lesueur, 'Pêche des aborigènes du Port Jackson', trans. J. Bonnemain, CL, MHN Le Havre, dossier 09 031.
62. For example, see C.-A. Lesueur, 'Grottes, chasse et pêche des sauvages du port Jackson, à la Nouvelle-Holland', plate 31 and Lesueur, 'Navigation', plate 34 in C.-A. Lesueur and N.-M. Petit, *Voyage de découvertes aux terres australes, Atlas Historique*, 2nd edn (Paris:

Arthus Bertrand, 1824). See also: B. Smith, *European Vision and the South Pacific, 1768–1850* (Oxford: Oxford University Press, 1960), pp. 147–8.

63. R. Jones, 'Images of Natural Man', pp. 52–7 and Morphy, 'Encountering Aborigines', pp. 152–3.

64. For example, see N.-M. Petit, 'Ourou-maré (dit *Bull-dog*), jeune guerrier de la tribu des Gwea-gal', plate 23, 'Norou-gal-derri, guerrier des environs de port Jackson, s'avançant pour combattre', plate 25 and 'Jeune femme sauvage de la tribu Bou-rou-bé-ron-gal, avec son enfant sur les épaules: Nouvelle-Galles du Sud', plate 28, in Lesueur and Petit, *Voyage de découvertes, Atlas Historique*.

65. Cuvier, 'Note Instructive', in *Aux origines de l'anthropologie*, pp. 174–5.

66. P. Jones, 'In the Mirror of Contact: Art of the French Encounters', in S. Thomas (ed.), *The Encounter 1802: Art of the Flinders and Baudin Voyages* (Adelaide: Art Gallery of South Australia, 2002), pp. 164–83, on p. 170.

67. See for instance, J. Neagle, 'Ben-nil-long' (London: Cadell and Davies, 1798), at http://nla.pic-an7566576-v [accessed 2 October 2012]; S. J. Neele, 'Benelong. A Native of New Holland' (London: s.n., *c.* 1790), at http://nla.gov.au/nla.pic-an9353133 [accessed 2 October 2012]; A. Earle, 'Portrait of Bungaree, a Native of New South Wales, with Fort Maquarie, Sydney Harbour, in Background' (*c.* 1826), at http://nla.gov.au/nla.pic-an2256865 [accessed 2 October 2012].

68. See, for example: M. Rosenthal, 'The Penitentiary as Paradise', in Thomas and Losche (eds), *Double Vision*, pp. 103–30.

69. J. Ravenet, 'Preliminary Sketch of a Botany Bay Scene and 'Reception of the Officers at Botany Bay', in King (ed.), *The Secret History of the Convict Colony*, on pp. 71 and 142.

70. A. Pellion, 'N^lle Hollande: Port-Jackson. Sauvages des Montagnes-Bleues', plate 66; 'N^lle Hollande: Port-Jackson. Sauvages des Environs de la Rivière Nepean', plate 100; 'N^lle Hollande: Port-Jackson. Sauvages des Montagnes-Bleues', plate 101; and J. Arago, 'N^lle Hollande: Port-Jackson. Sauvages des Environs de Sydney', plate 105, in J. Arago and A. Pellion, *Voyage autour du monde entrepris par ordre du roi, exécuté par les corvettes l'Uranie et la Physicienne pendant les années 1817, 1818, 1819 et 1820, Atlas Historique* (Paris: Pillet Aîné, 1825).

71. King (ed.), *The Secret History of the Convict Colony*, p. 3 and Dunmore, *French Explorers in the Pacific*, vol. 2, p. 40.

72. M.-S. Rivière, 'Distant Echoes of the Enlightenment: Private and Public Observations of Convict Life by Baudin's Disgraced Officer, Hyacinthe de Bougainville (1825)', *Australian Journal of French Studies*, 41:2 (2004), pp. 170–85.

6 Swans, Frogs and Rum: Natural History in an 'Unnatural' Space

1. 'La Nouvelle Hollande, cette contrée si vaste et si nouvelle encore pour le naturaliste, paraissait devoir nous fournir des collections et plus nombreuses et plus importantes, et cependant elles ont été presque nulles encore jusqu'à ce jour … heureusement notre relâche actuelle au Port jackson doit me mettre à même d'observer plus particulièrement les animaux de ce continent et je ne négligerais rien pour en tirer le plus grand avantage possible sous ce rapport'. F. Péron, *Tableau général d'une partie des espèces observées dans les diverses classes du Regne Animal parle Cen Fs Péron éléve-zoologiste attaché à l'expédition française de découvertes commandé par le Cen Nlas. Baudin*, CL, MHN Le Havre, dossier 21 003.

2. 'Rapport sur le voyage entrepris par les ordres du gouvernement et sous la direction de l'Institut, par le Capitaine Baudin', dated 26 December 1800, MNHN, ms 1214/6.

3. 'Plan of Itinerary for Citizen Baudin', p. 3.

4. B. Latour, *Science in Action: How to Follow Scientists and Engineers through Society* (Cambridge, MA: Harvard University Press, 1988), pp. 223–8.

5. M.-N. Bourguet, 'La collecte du monde: voyage et histoire naturelle (fin XVIIème siècle – début XIXème siècle)', in C. Blanckaert, C. Cohen, P. Corsi and J.-L. Fischer (eds), *Le Muséum au premier siècle de son histoire* (Paris: Éditions du Muséum National d'Histoire Naturelle, 1997), pp. 163–96.

6. The County of Cumberland (Sydney region) stretches from Broken Bay to the north, the Hawkesbury River to the north-west, the Nepean River to the west, the Cataract River to the south-west and the northern suburbs of Wollongong to the south.

7. Quoted in Outram, 'New Spaces in Natural History', pp. 259–61.

8. On the concept of the centre and the periphery in the production of knowledge, see Latour, *Science in Action*, pp. 219–28 and Bourguet, 'La collecte du monde', pp. 163–96.

9. 'Plan of Itinerary for Citizen Baudin', p. 1.

10. Y. Laisseau, 'Des Savants, pour quoi faire?', in Y. Laisseau (ed.), *Il y 200 ans:Les savants en Égypte* (Paris: Muséum National d'Histoire Naturelle and Éditions Nathan, 1998), p. 16.

11. J.-M. Drouin, 'Calculs et circonstances: portée et limites de l'œuvre des savants', in Laisseau (ed.), *Les savants en Égypte*, p. 91.

12. Tulard, *Napoleon: The Myth of the Saviour*, pp. 67 and 69.

13. Ibid., p. 69.

14. E. Said, *Orientalism* (New York: Vintage Books, 1979), p. 84.

15. 'l'expédition d'Égypte est ... aboutissement plutôt que rupture'. M.-N. Bourguet, 'Mission savantes au siècles des Lumières: du voyage à l'expédition', in Laisseau (ed.), *Les savants en Égypte*, p. 52.

16. Péron, *Tableau général*, CL, MHN Le Havre, 21 003.

17. Péron, *Voyage of Discovery*, vol. 1, p. 329.

18. Péron entitled one his notebooks 'Observations zoologiques de Port Jackson à la Nouvelle-Hollande' (CL, MHN Le Havre, dossier 21 001); however, this notebook only includes the catalogues relating to the second campaign. It would therefore appear that Péron intended also to describe or list the Port Jackson collection.

19. Péron's zoological collections from the Port Jackson sojourn and subsequent campaign are referred to in the two following notebooks: F. Péron, *Diarium zoographicum. No.XV. Ans XI et XII. Observationes generales de Collectionibus factis in Zoologiâ ex prefecturâ notrâ Portû Jackson (27 Brumaire an XI) adusque Promontorium Monoe Spei inclusivè (30 Pluviose an XIIe). Péron, zoologiste*, CL, MHN Le Havre, dossier 21 001; F. Péron, *Suite du Catalogue général des Descriptions, dessins, journaux et collections remis au Cen Baudin commandt en chef de l'expédition en exécution de l'ordre du Ministre de la Marine par le Cen Péron zoologiste à bord du Géographe*, CL, MHN Le Havre, dossier 21 002.

20. None of the zoological specimens collected at Port Jackson is included in the list of drawings made during the expedition. See F. Péron, *Tableau général de tous les dessins d'histoire Naturelle exécutés par M. Lesueur depuis notre départ de l'île de France pour les côtes de la Nouvelle-Hollande jusqu'à notre atterrissement sur les côtes de France*, CL, MHN Le Havre, dossier 21 001; F. Péron, *Journal No.VIII, Dessins et Plans, Géographico-zoologiques, Tableau Général de tous les dessins zoologiques exécutés par M. Lesueur dans chacune des diverses branches du Regne Animal distribués par classes avec le No particulier*

de la description des divers objets auxquels chacun d'eux appartient, CL, MHN Le Havre, dossier 21 002.

21. See M. Nugent, *Botany Bay: Where Histories Meet* (Crows Nest: Allen & Unwin, 2005), pp. 95–105.

22. Péron, *Voyage of Discovery*, vol. 1, p. 301.

23. Ibid.

24. F. Péron, 'Coquille donnée par Mme Paterson', CL, MHN Le Havre, dossier 21 013.

25. Outram, 'New Spaces in Natural History', p. 261.

26. Péron, *Voyage of Discovery*, vol. 1, p. 319; A.-L. de Jussieu, 'Notice sur l'expédition à la Nouvelle-Hollande, entreprise pour des recherches de Géographie et d'Histoire naturelle', *Annales du Muséum national d'histoire naturelle*, Paris, chez Levrault, Schœll et Compagnie, an XII (1804), vol. 5, p. 10.

27. Péron, *Voyage of Discovery*, vol. 1, pp. 320–1. See also Péron's notebooks in the dossier 78 and notes entitled '3ᵉ et 4ᵉ Classes, Quadrupède-ovipares et Reptiles', CL, MHN Le Havre, dossier 21 003–3.

28. See Péron's inventories of the zoological collection held in CL, MHN Le Havre, dossier 21, notebooks catalogued 21 001 and 21 002.

29. 'nous procurer les moyens de donner à nos recherches tout le développement possible'. Péron, *Voyage de découvertes*, vol. 2, p. 289; Péron, *Voyage of Discovery*, vol. 1, p. 302.

30. Péron, *Voyage of Discovery*, vol. 1, p. 312.

31. Ibid., p. 323.

32. Ibid., p. 331.

33. Péron, 'Conférence adressée à "Messieurs les Professeurs" décrivant les aborigènes et leurs mœurs près de Port Jackson', CL, MHN Le Havre, dossier 09 032, transcription J. Bonnemains.

34. Karskens, *The Colony*, pp. 258 and 260.

35. M. Sankey, 'French Representations of Sydney at the Beginning of the Nineteenth Century: the subversion of Modernism', *Literature of Aesthetics: The Journal of the Sydney Society of Literature and Aesthetics*, 15:2 (December 2005), pp. 101–8. See also Fornasiero and West-Sooby, 'Taming the Unknown', pp. 66.

36. Sankey, 'French Representations of Sydney', p. 105.

37. On the role that this concept played in the establishment and development of the British colony in Australia, see Gascoigne, *The Enlightenment and the Origins of European Australia*, pp. 86–7.

38. Leschenault, 'Account of the Vegetation of New Holland', p. 105.

39. Leschenault's comments were not approved of by first-lieutenant Louis Freycinet, who published the second edition of the *Voyage de découvertes*, in which this report appears. Based on the word of Governor King rather than on personal observation, Louis Freycinet contradicted Leschenault's claim that the colony was not harvesting enough grain to meet its needs. However, various other sources support Leschenault's statement by indicating that the colony was often short of provisions, including grain. Freycinet also pointed out that it was not only European plants that the British had introduced to Australia but also tropical plants. See the 'Account of the Vegetation of New Holland', pp. 108–9.

40. Leschenault, 'Account of the Vegetation of New Holland', p. 106.

41. Jangoux, 'Les zoologistes et botanistes qui accompagnèrent le capitaine Baudin', p. 66.

42. On Robert Brown's sojourn at Port Jackson, see E. W. Groves, D. T. Moore and T. G. Vallance (eds), *Nature's Investigator: The Diary of Robert Brown in Australia, 1801–1805*

(Canberra: Australia Biological Resources Study, 2001), pp. 201–16. For references to Leschenault in this text, see the entry dated Tuesday 11 May 1802, p. 203 as well as the letters from Robert Brown to Joseph Banks, Sydney, 30 May 1802, p. 206 and from Robert Brown to Jonas Dryander, Sydney, 30 May 1802, p. 207.

43. Arthur Phillip to Sir Joseph Banks, 16 November 1788, Papers of Sir Joseph Banks, Mitchell Library, http://www2.sl.nsw.gov.au/banks/series_37/37_view.cfm [accessed 5 March 2012].

44. W. Mayer, 'Deux géologues français en Nouvelle-Hollande (Australie): Louis Depuch et Charles Bailly, membres de l'expédition Baudin (1801–1803)', *Travaux du comité français d'histoire de la géologie*, 3rd series, 19:6 (meeting on 8 June 2005), pp. 95–112, on p. 107.

45. Péron, *Voyage de découvertes*, vol. 2, pp. 377–97; C. Bailly, 'Catalogue des objets de Minéralogie', CL, MHN Le Havre, dossier 21 004.

46. Quoted in Péron, *Voyage de découvertes*, vol. 2, p. 397.

47. Bailly, 'Catalogue des objets de Minéralogie', CL, MHN Le Havre, dossier 21 004.

48. Mayer, 'Deux géologues francais en Nouvelle-Hollande', p. 108.

49. Ibid., pp. 107–8; quoted in Péron, *Voyage de découvertes*, vol. 2, pp. 378–97.

50. See the account provided by Mayer in his article 'Deux géologues français en Nouvelle-Hollande', pp. 107–8 and Bailly's report, in Péron, *Voyage de découvertes*, vol. 2, pp. 378–97.

51. Philip Gidley King to Joseph Banks, Sydney, 5 June 1802, in Bladen (ed.), *HRNSW*, vol 4, pp. 782–6, on p. 784.

52. Bailly, in his 'Caisse no.1er', Catalogue des objets de Minéralogie appartenant au gouvernement qui m'ont été remis par le Cⁿ. Péron', CL, MHN Le Havre, dossier 21 004, lists forty-five samples, from five different types of rock: coal, schist, sandstone, breccia and quartz. Péron himself lists seven samples in his list entitled 'Minéraux', CL, MHN Le Havre, dossier 21 031.

53. 'On ne sera pas étonné que dans une recherche bornée à des côtes, la plupart désertes ou couvertes de bois, qui n'offroient ni montagnes élevées, ni ravins pour apercevoir les diverses couches de terre, ni aucun travail d'exploitation, les minéralogistes de Pusch et Bailly, n'aient pu recueillir qu'un petit nombre de minéraux insuffisans pour donner une idée exacte de la géologie de ce pays'. Jussieu, 'Notice sur l'expédition à la Nouvelle-Hollande, p. 7.

54. 'Plan du Port Jackson (Nouvelles Galles du Sud) d'après le Capitaine John Hunter assujetti aux observations faites à bord des Corvettes françaises en 1802', plate 29 in Freycinet, *Voyage de découvertes aux terres australes*.

55. 'Plan du Comté de Cumberland (Nouvelles Galles du Sud) d'après les Cartes Anglaises assujetti aux observations faites à bord des Corvette Françaises en 1802', plate 29 in Freycinet, *Voyage de découvertes, Navigation et géographie, Atlas*.

56. See 'Chart of the coast between Botany Bay and Broken Bay: surveyed in 1788 and 89 by Captain John Hunter' and 'A map of all those parts of the territory of New South Wales which have been seen by any person belonging to the settlement established at Port Jackson', in J. Hunter, *An Historical Journal of the Transactions at Port Jackson and Norfolk Island with the Discoveries which have been made in New South Wales and in the Southern Ocean since the Publication of Phillip's Voyage, Compiled from the Official Papers, including the Journals of Governors Phillip and King and of Lieut. Ball, and the Voyages from the first Sailing of the Sirius in 1787, to the Return of that Ship's Company to England in 1792* (Adelaide: Libraries Board of South Australia, 1968).

57. 'Plan de la ville de Sydney', plate 30 in Freycinet, *Voyage de découvertes, Navigation et géographie, Atlas*.

58. Péron, who gave consideration to various fields of natural history while at Port Jackson, did write about the geography of the region in the *Voyage of Discovery*, vol. 1, pp. 293–339.

59. These instructions are the same as those given to La Pérouse. See 'Private Instructions from the King to the Sieur de La Pérouse', pp. 42–7.

60. 'c'est certainement ce plan que Napoléon, dit-on, conservait sur son bureau'. J.-P. Faivre, *L'Expansion française dans le Pacifique, 1800–1842* (Paris: Nouvelles Editions Latines, 1953), p. 154.

61. *Nouvelle Hollande – Nouvelle Galles du Sud, 1° Chant, 2° Air de danse, 3° Cri de Ralliement*, notation faite par Lesueur et Bernier, plate 32, in Lesueur and Petit, *Voyage de découvertes, Atlas Historique*.

62. See C.-A. Lesueur, 'Nouvelle-Hollande, Nouvelle Galles du Sud, armes, ustencils et ornemens, plate 29, and 'Nouvelle-Hollande, Nouvelle Galles du Sud, dessins exécutés par les naturels', plate 33, in Lesueur and Petit, *Voyage de découvertes, Atlas Historique*.

63. P. P. King, *A Narrative of a Survey of the Intertropical and Western Coasts of Australia: Performed between 1818 and 1822. By Captain Phillip P. King. Vol. 1* (London: John Murray, 1827), p. 355.

64. F. Péron, 'Tableau no VIIe, Tableau des différents objets d'histoire naturelle remis au C^{en} Péron par différentes personnes', in 'Observations zoologiques de Port Jackson à la Nouvelle-Hollande', CL, MHN Le Havre, dossier 21 001 and and F. Péron, 'Inventaire général de tous les objets relatifs à l'histoire de l'homme, par François Péron', Copans and Jamin (eds), *Aux origines de l'anthropologie*, pp. 159–67.

65. F. Péron, *État des manuscrits confiés à M. Volney*, CL, MHN Le Havre, dossier 21 028.

66. Péron's inventory of the anthropological collection lists only three pieces collected from Tasmania, one from King George Sound and two earrings from elsewhere in Australia, whereas it shows that ten different types of objects in various quantities were acquired in Port Jackson. See Péron, 'Inventaire général', pp. 159–67.

67. L. F. Jauffret, 'Considerations to Serve in the Choice of Objects that May Assist in the Formation of the Special Museum of the *Société des Observateurs de l'Homme*, Requested of the Society by Captain Baudin', in Baudin, *The Journal*, pp. 594–6.

68. See the report written by Ransonnet for Baudin, dated 7 ventôse [26 February 1803], CL, MHN Le Havre, dossier 09 030.

69. N. Baudin, '1802 Commodore Rum Account', ANF, SM, 5JJ53.

70. N. Baudin, 'Compte général des dépenses relatives aux bâtiments de la République, le *Géograph*e, le *Naturaliste* et le *Casuarina* pendant la relâche au Port Jackson, Nouvelle-Hollande', ANF, SM, BB4997 and 'Commodore Rum Account', ANF, SM, 5JJ53; Andrew Thompson to Nicolas Baudin, Parramatta, 3 November 1802, ANF, SM, 5JJ53; H. Weld Noble to Nicolas Baudin, Sydney, 29 September 1802 ANF, SM, 5JJ53.

71. 'Je suis allé au-delà des lieux les plus avancés connus des Anglais'. Letter from Baudin to Jussieu, written at Port Jackson and dated 20 brumaire XI [11 November 1802], MNHN, ms 2082, part 5.

72. 'Le pin de la Nouvelle Zélande et le pin de la Norfolk Island, seront sans doute parmi les plantes vivantes, celles dont le prix sera mieux senti, puisqu'aucune nation européenne n'est encore parvenue à se les procurer'. Nicolas Baudin to the Minister of Marine and the Colonies, Port Jackson, 20 brumaire an X [11 November 1802].

73. Nicolas Baudin to Emmanuel Hamelin, Port Jackson, 26 brumaire an XI [17 November 1802], MNHN, ms 2082, part 9.
74. See A. Thouin, 'Extrait du Registre de délibérations de l'assemblée des Professeurs du Muséum d'histoire Naturelle', 10 messidor an XI [29 June 1803], ANF, F17, 3979, dossier 12.
75. See chapter 7 and J. Goy, *Les méduses de François Péron et de Charles-Alexandre Lesueur: un autre regard sur l'expédition Baudin* (Paris: Éditions du CTHS, 1995), p. 25.
76. Sankey, 'French Representations of Sydney', p. 104.

7 Baudin's 'New Expedition'

1. 'Je me recommande à votre souvenir et vais faire tous mes efforts pour compléter de nouveau une collection aussi nombreuse que celle que vous allez recevoir par le *Naturaliste*'. Nicolas Baudin to Antoine-Laurent de Jussieu, Port Jackson, 20 brumaire an XI [11 November 1802], MNHM, ms 2082, part 5.
2. Dunmore, *French Explorers in the Pacific*, vol. 2, p. 29.
3. Horner, *The French Reconnaissance*, p. 252.
4. 'Le commandant en était si pénétré, qu'il était bien décidé à son départ du Port Jackson d'y sacrifier tout le temps nécessaire etc. même sa vie pour remplir entièrement l'objet de sa mission'. Pierre-Bernard Milius to the Minister of Marine and the Colonies, Lorient, 4 germinal an XII [25 March 1804], ANF, SM, 5JJ24.
5. 'Plan of Itinerary for Citizen Baudin', pp. 1–6.
6. Nicolas Baudin to Antoine-Laurent de Jussieu, Port Jackson, 20 brumaire an XI [11 November 1802], MNHN, ms 2082, part 5 and Nicolas Baudin to the Minister of Marine and the Colonies, Port Jackson, 20 brumaire an XI [11 November 1802], ANF, SM, BB4 995.
7. 'pour la perfection de la géographie'. Nicolas Baudin to Antoine-Laurent de Jussieu, Port Jackson, 20 brumaire an XI [11 November 1802], MNHM, ms 2082, part 5.
8. 'Je ferai une nouvelle tentative pour les rencontrer [les isles du Romarin] afin de reprendre ensuite la terre de Witt, qui n'a pas la perfection nécessaire à la sûreté de la navigation'. Nicolas Baudin to Antoine-Laurent de Jussieu, Port Jackson, 20 brumaire an XI [11 November 1802], MNHM, ms 2082, part 5.
9. 'la terre de Leuwin, celles de la Concorde et de Witt, le canal d'Entrecasteaux, l'isle Maria et ses environs, la côte orientale de la grande isle de Diemen, les détroits de Basse et de Banks, et toute la côte sud-ouest de la Nouvelle Hollande depuis le promontoire de Wilson jusqu'aux isles St Pierre et St François ont été reconnues d'une manière suffisante pour la sûreté de la navigation'. Nicolas Baudin to Antoine-Laurent de Jussieu, Port Jackson, 20 brumaire an XI [11 November 1802], MNHM, ms 2082, part 5.
10. Fornasiero, Monteath and West-Sooby, *Encountering Terra Australis*, p. 383.
11. 'je crains que tant d'ouvrages ne prennent beaucoup plus de temps que ne le permettront les provisions que nous avons faites ici, car les observations géographiques exigent beaucoup de temps; et toutes reconnaissances faites trop promptement seront superficielles, imparfaites et remplies d'erreurs'. Nicolas Baudin to Antoine-Laurent de Jussieu, Port Jackson, 20 brumaire an XI [11 November 1802], MNHM, ms 2082, part 5.
12. Bourgueut, 'La collecte du monde', pp. 163–5 and Latour, *Science in Action*, p. 220.
13. Nicolas Baudin to the Minister of Marine and the Colonies, Port Jackson, 26 brumaire an XI [17 November 1802], ANF, SM, BB4995.

14. 'Nous avons eu le bonheur que sur la fin de notre séjour il est arrivé plusieurs bâtiments d'Angleterre qui ont considérablement diminué le prix des salaisons et des farines, ce qui nous a mis à même de nous compléter un an de vivres pour le *Casuarina* et pour moi, le *Naturaliste* n'en a eu que pour huit mois et ne doit faire aucune relâche'. Nicolas Baudin to the Minister of Marine and the Colonies, Port Jackson, 26 brumaire an XI [17 November 1802], ANF, SM, BB4995.

15. At Le Havre, the *Géographe* and the *Naturaliste* had been stocked with only eight months of provisions. See G. C. Ingleton, *Matthew Flinders: Navigator and Chartmaker* (Sydney: Genesis Publications Ltd, 1986), p. 70.

16. Baudin, *The Journal*, entry made 17 floréal an X [7 May 1802], p. 400.

17. Thomas Palmer to Nicolas Baudin, Sydney, 13 September 1802, ANF, SM, 5JJ53.

18. L. Freycinet, *Journal*, entry dated 26 vendémiaire an XI [18 October 1802], ANF, SM, 5JJ34.

19. T. Jamison, 'Supplied the French Corvette Géographe Commanded by Commodore Baudin on discovery with the following Medicines from His Majesty's Stores', compiled in Sydney and dated 2 November 1802, ANF, SM, 5JJ24.

20. Matthew Flinders to Nicolas Baudin, Port Jackson, undated, ANF, SM, 5JJ53.

21. 'Rapport sur le voyage', dated 26 December 1800, MNHN, ms 1214/6.

22. The number of crew members is stated by Horner, see *The French Reconnaissance*, p. 257.

23. The savants and artists were geographers Charles-Pierre Boullanger and Pierre Faure, zoologist François Péron, astronomer Pierre-François Bernier, mineralogist Charles Bailly, artists Nicolas-Martin Petit and Charles-Alexandre Lesueur, botanist Théodore Leschenault, junior gardener Antoine Guichenot, and pharmacist François Collas. Bailly, Leschenault, and Faure were transferred to the *Géographe* from the *Naturaliste*. For information and remarks about the staff changes see L. Freycinet, *Journal*, entry dated 12 brumaire an XI [3 November 1802]; Breton, *Journal*, entry dated 13 brumaire an XI [4 November 1802], ANF, SM, 5JJ57, and C. Baudin, *Souvenirs de jeunesse*, SHM, Vincennes: CC7a125, ms 116, p. 86. See also Horner, *The French Reconnaissance*, p. 257.

24. R. Sorrenson, 'The Ship as a Scientific Instrument in the Eighteenth Century', *Osiris*, 2nd series, 11(1996), pp. 221–36, on p. 222.

25. Horner, *The French Reconnaissance*, p. 57.

26. Nicolas Baudin to the members of the *Institut National*, 6 floréal an VIII [26 April 1800], ANF, SM, BB4995.

27. James Underwood and Henry Kable to Nicolas Baudin, dated 7 July 1802, ANF, SM, 5JJ53.

28. L. Freycinet, *Journal*, entry undated, ANF, SM, 5JJ49.

29. 'Je m'imaginais qu'il fournirait au capitaine de ce bâtiment, toutes les facilités et tous les moyens possibles de faire cet ouvrage avec exactitude'. L. Freycinet, *Journal*, entry undated.

30. 'Je m'estimerai heureux, si après les peines et les travaux du voyage que nous allons entreprendre, je puis avoir un peu contribué au succès de l'Expédition que vous commandez'. Demonstrating his enthusiasm for the objectives of the expedition, Louis Freycinet wrote to Baudin: 'I will be happy if after the difficulties and work of the voyage that we are going to undertake, I have contributed a little to the success of the expedition that you command'. Port Jackson, 2 vendémiaire an XI [24 September 1802], in L. Freycinet, *Journal*, entry undated.

31. 'L'acquisition que j'ai faite au nom du gouvernement français du petit bâtiment dont je vous ai confié le commandement; n'ayant d'autre but qu'une plus grande facilité dans nos

travaux, et une reconnaissance plus exacte des côtes que le défaut d'eau ou autres difficultés pourraient m'empêcher d'approcher d'assez près, vous devez apporter la plus grande attention à ne jamais vous écarter du *Géographe* soit de jour soit de nuit, que vous n'ayez reçu un ordre particulier à ce sujet. Dans toutes les circonstances, vous manœuvrerez de façon à ne jamais vous éloigner hors de la vue'. L. Freycinet, *Journal*, entry dated 24 and 25 brumaire an XI [15 and 16 November 1802].

32. 'Tant que vous serez avec le *Géographe* je vous fournirai tous les secours nécessaires'. Freycinet, *Journal*, entry dated 3 and 4 brumaire an XI [25 and 26 October 1802].

33. 'Avez-vous donc envie de vous séparer de moi?' L. Freycinet, *Journal*, entry dated 3 and 4 brumaire an XI [25 and 26 October 1802].

34. '[Brèvedent] aura votre table, mais il couchera en avant dans la cabine qui a été faite pour lui'. Letter from Nicolas Baudin to Louis Freycinet, written at Port Jackson and dated 1 vendémiaire an XI [23 September 1802], in L. Freycinet, *Journal*, entry undated.

35. 'je pourrai désormais tout examiner et ne rien laisser à faire à ceux qui nous succéderont dans un semblable travail'. Nicolas Baudin to Antoine-Laurent de Jussieu, Port Jackson, dated 20 brumaire an XI [11 November 1802], MNHM, ms 2082, part 5.

36. Nicolas Baudin to the members of the *Institut National*, 6 floréal an VIII [26 April 1800], ANF, SM, BB4995.

37. 'Plan of Itinerary for Citizen Baudin', p. 2.

38. Ibid., p. 3.

39. Baudin, *The Journal*, pp. 491–2.

40. Péron, 'Catalogue zoologique', CL, MHN Le Havre, dossier 21 002.

41. Horner, *The French Reconnaissance*, p. 358.

42. *Vombatus ursinus ursinus* was a subspecies of the common wombat (*Vombatus ursinus*), which was once found on several Bass Strait islands but now survives only on Flinders Island. See L. J. Pigott and L. Jessop, 'The Governor's Wombat: Early History of an Australian Marsupial', *Archives of Natural History*, 34:2 (2007), pp. 207–18, on p. 216.

43. Antoine-Jean-Marie Thévenard to the Minister of Marine and the Colonies, Lorient, dated 8 germinal an XII [29 March 1804], ANF, SM, BB4996.

44. Jussieu, 'Notice sur l'expédition à la Nouvelle-Hollande', p. 6.

45. Péron, 'Observations zoologiques de Port Jackson à la Nouvelle-Hollande', CL, MHN Le Havre, dossier 21 001.

46. Antoine-François Fourcroy to Pierre-Bernard Milius, Paris, dated 29 germinal an 12 [19 April 1804], Milius, *Récit du voyage*, p. 61.

47. Thouin, 'Extrait du Registre de déliberations', ANF, F17, 3979, dossier 12.

48. Nicolas Baudin to Emmanuel Hamelin, Port Jackson, dated 26 brumaire an XI [17 November 1802], MNHN, ms 2082, part 9.

49. 'cette perte vraiment déplorable'. Thouin, 'Extrait du Registre de déliberations', ANF, F17, 3979, dossier 12.

50. Thouin recorded that there were fourteen crates of mineralogical samples on the *Naturaliste*, whereas fragmentary records indicate that there may have been only five aboard the *Géographe*. See Thouin, 'Extrait du Registre de déliberations', ANF, F17, 3979, dossier 12; P.-B. Milius, 'Etat des objets d'histoire naturelle que j'ai apporté en France sur la corvette Le *Géographe* que je commandai et qui ont été recueillis en grande partie par le commandant', Milius, *Récit du voyage*, pp. 62–3; and Bailly, 'Caisse no.1er, Catalogue des objets de Minéralogie appartenant au gouvernement qui m'ont été remis par le C^n. Péron', CL, MHN Le Havre, dossier 21 004.

51. Baudin, *The Journal*, pp. 486–7.

52. See the report written by Ransonnet for Baudin, dated 7 ventôse [26 February 1803], CL, MNH Le Havre, dossier 09 030.
53. Baudin, *The Journal*, pp. 507–8.
54. Ibid., pp. 486–7.
55. Ibid., pp. 506, 511, 519, 522–3, 525 and 527.
56. 'Plan of Itinerary for Citizen Baudin', p. 4.
57. 'cette partie du voyage ne fut pas désastreuse comme la première'. Jussieu, 'Notice sur l'expédition à la Nouvelle-Hollande', vol. 5, p. 3.
58. Nicolas Baudin to the Minister of Marine and the Colonies, Kupang Bay and dated 9 Prairéal an XI [29 May 1803], ANF, SM, BB4995.

8 Epilogue: Voyaging into the Nineteenth Century

1. 'il me reste assez de force dans ce moment pour vous assurer que les intentions du Gouvernement sont remplies et ce voyage sera honorable pour les français'. Nicolas Baudin to the Minister of Marine and the Colonies, Port-Louis, 13 thermidor an XI [31 July 1803], ANF SM BB4995.
2. Dunmore, *French Explorers in the Pacific*, vol. 2, p. 386.
3. Ibid., p. 11.

WORKS CITED

Primary Sources

Manuscripts

Archives Nationales de France, Série Marine (ANF, SM), BB4 995–7 and 5JJ24–57.

Collection Lesueur, Muséum d'Histoire Naturelle, Le Havre (CL, MHN Le Havre).

Mitchell Library, Sydney (ML), James Hingston Tuckey – papers, 1804, microfilm – CY 1249, frames 58–122.

Mitchell Library, Sydney (ML), M. Flinders, 'Journal on the Investigator', January 1801–July 1802, vol. 1, safe 1/24.

Mitchell Library, Sydney (ML), Franklin family – Letters from John Franklin to family members, 1802–3, citation no. C 231.

Mitchell Library, Sydney (ML), King family – Correspondence and memoranda, 1775–1806, reference A 1980/2 CY 906.

Muséum National d'Histoire Naturelle (MNHN), Lettres de Capitaine Baudin, ms 2082.

Muséum National d'Histoire Naturelle (MNHN), 'Rapport sur le voyage entrepris par les ordres du gouvernement et sous la direction de l'institut, par le Capitaine Baudin', 26 December 1800, ms 1214/6.

National Library of Australia (NLA), Papers of Sir Joseph Banks 1745–1923, series 1: Correspondence and Papers of Sir Joseph Banks, ms 9/22.

National Maritime Museum, Greenwich, The Flinders Papers: letters and documents about the explorer Matthew Flinders (1774–1814), FLI07.

Service Historique de la Defence, Archives Centrale de la Marine (SHD, Marine), Dossiers Individuels du Personnel, CC7.

Printed Sources

Arago J. and A. Pellion, *Voyage autour du monde entrepris par ordre du roi, exécuté par les corvettes l'Uranie et la Physicienne pendant les années 1817, 1818, 1819 et 1820, Atlas Historique* (Paris: Pillet Aîné, 1825).

Baudin, N., *The Journal of Post-Captain Nicolas Baudin, Commander in Chief of the Corvettes* Géographe *and* Naturaliste, *Assigned by Order of the Government to a Voyage of Discovery*, trans. C. Cornell (Adelaide: Libraries Board of South Australia, 1974).

—, *Mon voyage aux terres australes: journal personnel du commandant Baudin, illustré par Lesueur et Petit*, ed. J. Bonnemains (Paris: Imprimérie Nationale, 2000).

—, *Journal du voyage aux Antilles de* La Belle Angélique, *1796–1798*, ed. M. Jangoux (Paris: Press de l'Université Paris-Sorbonne, 2009).

Bladen, F. M. (ed.), *Historical Records of New South Wales*, vols 4 and 5 (Sydney: Government Printer, 1896–7).

Chambers, N. (ed.), *The Letters of Sir Joseph Banks: A Selection, 1768–1820* (London: Imperial College Press, 2000).

Collins, D., *An Account of the British Colony in New South Wales, with Remarks on the Dispositions, Customs, Manners, etc. of the Native Inhabitants of that Country*, vols 1 and 2 (London: A. H. & A. W. Reed, 1802).

Copans, J. and J. Jamin (eds), *Aux origines de l'anthropologie française: Les mémoires de la société des observateurs de l'homme en l'an VIII* (Paris: Le Sycomore, 1978).

De Galaup de La Pérouse, J.-F., *A Voyage Round the World, Performed in the Years 1785, 1786, 1787 and 1788 by the* Boussole *and the* Astrolabe, *Under the Command of J. F. G. De la Pérouse*, trans. L. A. Millet-Mureau (London: S. Hamilton, 1799).

Degérando, J.-M., 'Considérations sur les diverses méthodes à suivre dans l'observation des peuples sauvages', in J. Copans, and J. Jamin (eds), *Aux origines de l'anthropologie française: Les mémoires de la société des observateurs de l'homme en l'an VIII* (Paris: Le Sycomore, 1978), pp. 127–69.

de Jussieu, A.-L., 'Notice sur l'expédition à la Nouvelle-Hollande, entreprise pour des recherches de Géographie et d'Histoire naturelle', *Annales du Muséum national d'histoire naturelle*, Paris, chez Levrault, Schœll et Compagnie, an XII (1804), vol. 5, p. 10.

De la Roquette, D., 'Notice historiques sur MM. Henri et Louis Freycinet', *Bulletin de la Société de Géographie*, 20:2 (1843), pp. 501–37.

Freycinet, L., *Voyage de découvertes aux terres australes: exécuté par ordre de sa majesté l'empereur et roi, sur les corvettes le* Géographe, *le* Naturaliste; *et la goëlette le* Casuarina, *pendant les années 1800, 1801, 1802, et 1804, Navigation et géographie, Atlas* (Paris: Imprimerie Royale, 1812).

—, *Voyage autour du monde, entrepris par ordre du roi, sous le ministère et conformément aux instruction de S. Exc. M. le Vicomte du Bouchage, secrétaire d'état au département de la marine, exécuté sur les corvettes de S. M. l'*Uranie *et la* Physicienne, *pendant les années 1817, 1818, 1819 et 1820*, vol. 2, *Historique* (Paris: Pillet Aîné, 1839).

Gazette Nationale ou Le Moniteur Universel du 29 Vendémiaire an IX [21 October 1800], Collection Lesueur, Muséum d'histoire naturelle du Havre (hereafter CL, MHN), dossier 06 002.

Hunter, J., *An Historical Journal of the Transactions at Port Jackson and Norfolk Island with the Discoveries which have been made in New South Wales and in the Southern Ocean since the Publication of Phillip's Voyage, Compiled from the Official Papers, including the Journals of Governors Phillip and King and of Lieut. Ball, and the Voyages from the first Sailing of the Sirius in 1787, to the Return of that Ship's Company to England in 1792* (Adelaide: Libraries Board of South Australia, 1968).

Jauffret, L.-F., 'Introduction aux mémoires de la Société des Observateurs de l'Homme, lu dans la séance du 18 messidor an IX', in J. Copans, and J. Jamin (eds), *Aux origines de l'anthropologie française: Les mémoires de la société des observateurs de l'homme en l'an VIII* (Paris: Le Sycomore, 1978), pp. 73–85.

King, P. P., *A Narrative of a Survey of the Intertropical and Western Coasts of Australia: Performed between 1818 and 1822. By Captain Phillip P. King. Vol. 1* (London: John Murray, 1827).

Leschenault, T., 'Account of the Vegetation of New Holland and Van Diemen's Land', in F. Péron [L. Freycinet], *Voyage of Discovery to the Southern Lands*, vol. 3, *Dissertations on Various Subjects*, trans. C. Cornell (Adelaide: The Friends of the State Library of South Australia, 2007), pp. 97–109.

Malaspina, A., 'Loose Notes on the English Colony of Port Jackson', in R. J. King (ed.), *The Secret History of the Convict Colony: Alexandro Malaspina's Report on the British Settlement of New South Wales* (Sydney: Allen & Unwin, 1990), pp. 132–50.

Milius, P.-B., 'Coup d'oeil rapide sur l'établissement des Anglais à la Nouvelle Hollande', Bibliothèque Municipale de Caen, Archive du général Decaen (hereafter BC, AGD) vol. 92, ms 177, fols 74–8 at http://sydney.edu.au/arts/research/baudin/written_records/other_documents.shtml [accessed 2 January 2012].

—, *Récit du voyage aux terres australes par Pierre-Bernard Milius, second sur le* Naturaliste *dans l'expédition Baudin (1800–1804)*, ed. J. Bonnemains and P. Haugel (Le Havre: Société havraise d'études diverses, 1987).

—, *Récit du voyage aux Terres australes de Pierre-Bernard Milius, second sur le* Naturaliste *dans l'expédition Baudin (1800–1804)*, ed. J. Bonnemains and P. Haugel (Le Havre: Société havraise d'études diverses, 2000).

Péron, F., *Voyage de découvertes aux terres australes exécuté par ordre de sa majesté l'empereur et roi, sur les corvettes le* Géographe, *le* Naturaliste; *et la goélette le* Casuarina, *pendant les années 1800, 1801, 1802, 1803, et 1804*, vol. 1, *Historique* (Paris: Imprimerie Imperiale, 1807).

— [and L. Freycinet], *Voyage de découvertes aux terres australes: exécuté par ordre de sa majesté l'empereur et roi, sur les corvettes le* Géographe, *le* Naturaliste; *et la goélette le* Casuarina, *pendant les années 1800, 1801, 1802, et 1804, Historique, Atlas par MM. Lesueur et Petit*, 2nd edn (Paris: Arthus Bertrand, 1824).

—, 'Rapport de François Péron au Général Decaen sur les Colonies anglaise de la Nouvelle Hollande', BC, AGD, vol. 92, fol. 2, at http://sydney.edu.au/arts/research/baudin/written_records/other_documents.shtml [accessed 2 January 2012].

—, [L. De Freycinet], *Voyage of Discovery to the Southern Lands*, trans. C. Cornell, vol. 1, 2nd edn (1824) (Adelaide: The Friends of the State Library of South Australia, 2006).

'Rapport et projet de décret sur le mode d'épurement de la marine civile et militarie, présenté, au nom du Comité de marine, à la Convention nationale, par Topsent, député du Département de l'Eure'; Archives Parlementaires de 1787 à 1860 (Paris: Librairie Administrative Paul Dupont, 1910).

Ravenet, J., 'Preliminary Sketch of a Botany Bay Scene and Reception of the Officers at Botany Bay', in R. J. King (ed.), *The Secret History of the Convict Colony: Alexandro Malaspina's Report on the British Settlement of New South Wales* (Sydney: Allen & Unwin, 1990).

Rivière, M.-S., *A Woman of Courage: The Journal of Rose de Freycinet on her Voyage Around the World, 1817–1820* (Canberra: National Library of Australia, 1996).

Watson, F. (ed.), *Historical Records of Australia*, series 1, vol. 4 (Sydney: Library Committee of the Commonwealth Parliament, 1915).

Secondary Sources

Atkinson, A., *The Europeans in Australia: A History*, vol.1 (Oxford: Oxford University Press, 1997).

Blainey, G., *The Tyranny of Distance: How Distance Shaped Australia's History* (Melbourne: Sun Books, 1966).

Blanckaert, C., 'L'anthropologie en France, le mot et l'histoire (XIVe–XIXe siècles)', *Bulletins et Mémoires de la Société d'anthropologie de Paris*, new series, 1:3–4 (1989), pp. 20–2.

—, '1800 – Le moment "naturaliste" des sciences de l'homme', *Revue d'Histoire des Sciences Humaines*, 3 (2000), pp. 117–60.

Blanckaert,C., C. Cohen, P. Corsi and J.-L. Fischer (eds), *Le Muséum au premier siècle de son histoire* (Paris: Éditions du Muséum national d'histoire naturelle, 1997).

Bonnemains, J., E. Forsyth and B. Smith (eds), *Baudin in Australian Waters: The Artwork of the French Voyage of Discovery to the Southern Lands 1800–1804* (Oxford and Melbourne: Oxford University Press, 1988).

Bourguet, M.-N., 'Race et folklore: L'image officielle de la France en 1800', *Annales. Histoire, Sciences Sociales*, 31:4 (July–August 1976), pp. 802–23.

—, 'La collecte du monde: voyage et histoire naturelle (fin XVIIème siècle – début XIXème siècle)', in C. Blanckaert, C. Cohen, P. Corsi and J.-L. Fischer (eds), *Le Muséum au premier siècle de son histoire* (Paris: Éditions du Muséum national d'Histoire naturelle, 1997), pp. 163–96.

Brown, A., *Ill-Starred Captains: Flinders and Baudin* (Fremantle: Fremantle Press, 2004).

Chappey, J.-L., *La société des observateurs de l'homme: des anthropologues au temps de Bonaparte* (Paris: Société des études robespierristes, 2002).

Clark, C. M. H., *A History of Australia*, vol. 1, *From the Earliest Times to the Age of Macquarie* (Melbourne: Melboure University Press, 1962).

Conley, M. A., *From Jack Tar to Union Jack: Representing Naval Manbood in the British Empire, 1870–1918* (Manchester: Manchester University Press, 2009).

Cormack, W. S., *Revolution and Political Conflict in the French Navy, 1789–1794* (Cambridge: Cambridge University Press, 1995).

Dear, I. C. B. and P. Kemp, *The Oxford Companion to Ships and the Sea* (Oxford: Oxford University Press, 1976).

De Certeau, M., D. Julia and J. Revel, *Une politique de la langue: la révolution française et les patois – l'enquête de Grégoire* (Paris: Gallimard, 1975).

De Beer, G., *The Sciences Were Never at War* (London: Thomas Nelson and Sons, 1960).

Dening, G., *Mr Bligh's Bad Language: Passion, Power and Theatre on the Bounty* (Cambridge: Cambridge University Press, 1992).

—, *Performances* (Melbourne: Melbourne University Press, 1996).

Douglas, B., 'Climate to Crania: Science and the Racialization of Human Difference, 1750–1880', in B. Douglas and C. Ballard (eds), *Foreign Bodies: Oceania and the Science of Race, 1750–1940* (Canberra: ANU E-Press, 2008), pp. 33–96.

Dunmore, J., *French Explorers in the Pacific, vol. 2: The Nineteenth Century* (Oxford: Oxford University Press, 1969).

Duyker, E., *François Péron: An Impetuous Life* (Melbourne: Miegunyah Press, 2006).

Dwyer, P., *Napoleon: The Path to Power, 1769–1799* (London: Bloomsbury, 2007).

Faivre, J.-P., *L'expansion française dans le Pacifique, 1800–1842* (Paris: Nouvelles Éditions Latines, 1953).

Fornasiero, J., 'Deux observateurs de l'homme aux antipodes: Nicolas Baudin et François Péron', in M. Jangoux (ed.), *Portés par l'air du temps: les voyages du Capitaine Baudin*, special number of *Études sur le 18ᵉᵐᵉ siècle*, 38 (2010), pp. 157–70.

Fornasiero, J. and J. West-Sooby, 'Taming the Unknown: The Representation of Terra Australis by the Baudin Expedition, 1801–1803', in A. Chittleborough, G. Dooley, B. Glover and R. Hosking, *Alas for the Pelicans: Flinders, Baudin and Beyond* (Kent Town: Wakefield Press, 2002), pp. 59–80.

—, 'A Cordial Encounter? The Meeting of Matthew Flinders and Nicolas Baudin (8–9 April 1802)', in I. Coller, H. Davies and J. Kalmann (eds), *History and Civilization: Papers from the George Rudé Seminar* (2005), vol. 1, pp. 53–61.

—, *French Designs on Colonial New South Wales: François Péron's Memoir on the English Settlements in New Holland, Van Diemen's Land and the Archipelagos of the Great Pacific Ocean* (Adelaide: Friends of the State Library of South Australia, 2013).

Fornasiero, J., P. Monteath and J. West-Sooby, *Encountering Terra Australis: The Australian Voyages of Nicolas Baudin and Matthew Flinders* (Kent Town: Wakefield Press, 2004).

Forth, C. E. (ed.), *Masculinity in the Modern West: Gender, Civilization and the Body* (Basingstoke: Palgrave Macmillan, 2008).

Garrioch, D., *The Making of Revolutionary Paris* (Berkeley, CA: University of California Press, 2002).

Gascoigne, J., *Science in the Service of Empire: Joseph Banks, the British State and the Uses of Science in the Age of Revolution* (Cambridge: Cambridge University Press, 1998).

—, *The Enlightenment and the Origins of European Australia* (Cambridge: Cambridge University Press, 2002).

—, *Captain Cook: Voyager between Worlds* (London: Continuum Books, 2007).

Godechot, J., B. F. Hyslop and D. L. Dowd, *The Napoleonic Era in Europe* (New York: Holt, Rinehart and Winston, 1971).

Goy, J., *Les méduses de François Péron et de Charles-Alexandre Lesueur: un autre regard sur l'expédition Baudin* (Paris: Éditions du CTHS, 1995).

Groves, E. W., D. T. Moore and T. G. Vallance (eds), *Nature's Investigator: The Diary of Robert Brown in Australia, 1801–1805* (Canberra: Australia Biological Resources Study, 2001).

Harrison, C. E., 'Projections of the Revolutionary Nation: French Expeditions in the Pacific, 1791–1803', *Osiris*, 24 (2009), pp. 33–54.

—, 'Replotting the Ethnographic Romance: Revolutionary Frenchmen in the Pacific, 1769–1804', *Journal of the History of Sexuality*, 21:1 (January 2012), pp. 39–59.

Horner, F., *The French Reconnaissance: Baudin in Australia, 1801–1803* (Carlton: Melbourne University Press, 1987).

—, *Looking for La Pérouse: D'Entrecasteaux in Australia and the South Pacific, 1792–1793* (Melbourne: Melbourne University Press, 1995).

Hufton, O., *Women and the Limits of Citizenship in the French Revolution* (Toronto: University of Toronto Press, 1999).

Hughes, M. J., 'Philosphical Travellers at the Ends of the Earth: Baudin, Péron and the Tasmanians', in R. W. Home (ed.), *Australian Science in the Making* (Cambridge: Cambridge University Press, 1998), pp. 23–44.

—, 'Making Frenchmen into Warriors: Martial Masculinity in Napoleonic France', in C. E. Forth and B. Taithe (eds), *French Masculinities: History, Culture and Politics* (Basingstoke: Palgrave Macmillan, 2007), pp. 53–4.

Hunt S. and P. Carter, *Terre Napoléon: Australia through French Eyes, 1800–1804* (Sydney: Historic Houses Trust of New South Wales, 1999).

Ingleton, G. C., *Matthew Flinders: Navigator and Chartmaker* (Sydney: Genesis Publications Ltd, 1986).

Jacob, M., *Living the Enlightenment: Freemasonry and Politics in Eighteenth-Century Europe* (Oxford: Oxford University Press, 1991).

Jangoux, M., 'La première relâche du *Naturaliste* au Port Jackson (26 avril–18 mai 1802): le témoignage du capitaine Hamelin', *Australian Journal of French Studies*, 41:2 (2004), pp. 126–51.

—, 'Les zoologistes et botanistes qui accompagnèrent le capitaine Baudin aux terres australes', *Australian Journal of French Studies*, 41:2 (2004), pp. 55–78.

Jones, C., *The Great Nation: France from Louis XI to Napoleon* (London: Penguin, 2003).

—, *Paris: Biography of a City* (London: Penguin, 2004).

Jones, P., 'In the Mirror of Contact: Art of the French Encounters', in S. Thomas (ed.), *The Encounter 1802: Art of the Flinders and Baudin Voyages* (Adelaide: Art Gallery of South Australia, 2002), pp. 164–83.

Jones, R., 'Images of Natural Man', in J. Bonnemains, E. Forsyth and B. Smith (eds), *Baudin in Australian Waters: The Artwork of the French Voyage of Discovery to the Southern Lands 1800–1804* (Oxford and Melbourne: Oxford University Press, 1988), pp. 35–64.

Karskens, G., *The Colony: A History of Early Sydney* (Crows Nest: Allen & Unwin, 2009).

Konishi, S., *The Aboriginal Male in the Enlightenment World* (London: Pickering & Chatto, 2012).

Laisseau, Y. (ed.), *Il y 200 ans: Les savants en Égypte* (Paris: Muséum National d'Histoire Naturelle and Éditions Nathan, 1998).

Lamb, J., V. Smith and N. Thomas, *Exploration and Exchange: A South Seas Anthology, 1689–1900* (Chicago, IL: University of Chicago Press, 2000).

Latour, B., *Science in Action: How to Follow Scientists and Engineers through Society* (Cambridge, MA: Harvard University Press, 1988).

Lyons, M., *Napoleon Bonaparte and the Legacy of the French Revolution* (New York: St Martin's Press, 1994).

Martin, B. J., *Napoleonic Friendship: Military Fraternity, Intimacy and Sexuality in Nineteenth-Century France* (Durham, NH: University of New Hampshire: University Press of New England, 2011).

Mayer, W., 'Deux géologues français en Nouvelle-Hollande (Australie): Louis Depuch et Charles Bailly, membres de l'expédition Baudin (1801–1803)', *Travaux du comité français d'histoire de la géologie*, 3rd series, 19:6 (meeting on 8 June 2005), pp. 95–112.

McKellar, M. (ed.), *Strangers in a Foreign Land: The Journal of Niel Black and Other Voices from the Western District* (Carlton: Miegunyah Press, 2008).

McPhee, P., *A Social History of France, 1789–1914* (Basingstoke: Palgrave Macmillan, 2004).

Morphy, H., 'Encountering Aborigines', in S. Thomas (ed.), *The Encounter 1802: Art of the Flinders and Baudin Voyages* (Adelaide: Art Gallery of South Australia, 2002), pp. 148–63.

Nugent, M., *Botany Bay: Where Histories Meet* (Crows Nest: Allen & Unwin, 2005).

Outram, D., 'New Spaces in Natural History', in N. Jardine, J. A. Secord and E. C. Spary (eds), *Cultures of Natural History* (Cambridge: Cambridge University Press, 1996), pp. 249–65.

—, *The Enlightenment*, 2nd edn (Cambridge: Cambridge University Press, 2005).

Pigott, L. J. and L. Jessop, 'The Governor's Wombat: Early History of an Australian Marsupial', *Archives of Natural History*, 34:2 (2007), pp. 207–18.

Reddy, W., 'Sentimentalism and its Erasure: The Role of Emotions in the Era of the French Revolution', *Journal of Modern History*, 72:1 (March 2000), pp. 109–52.

—, *The Navigation of Feeling: A Framework for the History of Emotions* (Cambridge: Cambridge University Press, 2001).

Rigondet, G., *François Péron 1775–1810 et l'expédition du commandant Nicolas Baudin: les Français à la découverte de l'Australie* (Charroux: Éditions des Cahiers Bourbonnais, 2002).

Rivière, L. P., 'Un périple en Nouvelle Hollande au début du XIXe siècle', *Comptes-rendus mensuels des séances de l'Académie des Sciences coloniales*, 13 (1953), pp. 571–89.

Rivière, M.-S., 'Distant Echoes of the Enlightenment: Private and Public Observations of Convict Life by Baudin's Disgraced Officer, Hyacinthe de Bougainville (1825)', *Australian Journal of French Studies*, 41:2 (2004), pp.170–85.

Rodger, N. A. M., *The Wooden World: An Anatomy of the Georgian Navy* (London: Collins, 1986).

Rosenthal, M., 'The Penitentiary as Paradise', in N. Thomas and D. Losche (eds), *Double Vision: Art Histories and Colonial Histories in the Pacific* (Cambridge: Cambridge University Press, 1999), pp. 103–30.

Rosenwein, B., *Emotional Communities in the Early Middle Ages* (Ithaca, NY: Cornell University Press, 2006).

Rudé, G., *Revolutionary Europe, 1783–1815* (Glasgow: Fontana, 1964).

Rusden, G. W., *History of Australia*, vol. 1 (Cambridge: Cambridge University Press, 1883), pp. 301–2.

Russell, P., *Savage or Civilized? Manners in Colonial Australia* (Sydney: University of New South Wales Press, 2010).

Said, E., *Orientalism* (New York: Vintage Books, 1979).

Sankey, M., 'The Baudin Expedition in Port Jackson, 1802: Cultural Encounters and Enlightenment Politics', *Explorations*, 31 (December 2001), pp. 5–36.

—, 'The Aborigines of Port Jackson, as seen by the Baudin Expedition', *Australian Journal of French Studies*, 41:2 (2004). pp. 117–51.

—, 'French Representations of Sydney at the Beginning of the Nineteenth Century: the Subversion of Modernism', *Literature of Aesthetics: The Journal of the Sydney Society of Literature and Aesthetics*, 15:2 (December 2005), pp. 101–8.

Scott, E., *Terre Napoleon: A History of French Explorations and Projects in Australia* (London: Taylor and Francis, 1910).

Smith, B., *European Vision and the South Pacific, 1768–1850* (Oxford: Oxford University Press, 1960).

Sorrenson, R., 'The Ship as a Scientific Instrument in the Eighteenth Century', *Osiris*, 2nd series, 11(1996), pp. 221–36.

Southwood, J. and D. Simpson, 'Baudin's Doctors: French Medical Scientists in Australian Waters, 1801–1803', *Australian Journal of French Studies*, 41:2 (2004), pp. 152–64.

Starbuck, N., 'Sir Joseph Banks and the Baudin Expedition: Exploring the Politics of the Republic of Letters', in G. Betros (ed.), *French History and Civilization: Papers from the George Rudé Seminar*, vol. 3 (2009), pp. 56–68.

—, 'The Colonial Field: Science, Sydney and the Baudin Expedition (1802)', *Explorations*, 52 (June 2012), pp. 3–35.

Staum, M., *Minerva's Message: Stabilizing the French Revolution* (Montreal: McGill-Queen's Press, 1996).

—, 'The Paris Geographical Society Constructs the Other, 1821–1850', *Journal of Historical Geography*, 26:2 (2000), pp. 222–38.

Stocking, G., 'French Anthropology in 1800', *Isis*, 55:180 (1964), pp. 134–50.

Thomas, N., *In Oceania: Visions, Artifacts, Histories* (London: Duke University Press, 1997).

Thomas, N. and D. Losche (eds), *Double Vision: Art Histories and Colonial Histories in the Pacific* (Cambridge: Cambridge University Press, 1999).

Tulard, J., *Napoleon: The Myth of the Saviour*, trans. T. Waugh (London: Methuen and Co., 1985).

Webb, J., *George Caley: Nineteenth Century Naturalist* (Chipping Norton: Surrey Beatty and Sons, 1995).

West-Sooby, J., 'Une expédition sous haute surveillance: le voyage aux terres australes vu par les Anglais', in M. Jangoux (ed.), *Portés par l'air du temps: les voyages du Capitaine Baudin*, special number of *Études sur le 18ème siècle*, 38 (2010), pp. 187–201.

Wolf, S., 'French Civilization and Ethnicity in the Napoleonic Empire', *Past and Present*, 124 (August 1989), pp. 96–120.

INDEX